Ben Stacy Jerrik (Ed.)

Arkansas Highway 204

Ben Stacy Jerrik (Ed.)

Arkansas Highway 204

Siloam Springs, Arkansas State Highway and Transportation Department, Arkansas

Part Press

Imprint
Permission is granted to copy, distribute and/or modify this document under the terms of the GNU Free Documentation License, Version 1.2 or any later version published by the Free Software Foundation; with no Invariant Sections, with the Front-Cover Texts, and with the Back- Cover Texts. A copy of the license is included in the section entitled "GNU Free Documentation License".

All parts of this book are extracted from Wikipedia, the free encyclopedia (www.wikipedia.org).

You can get detailed informations about the authors of this collection of articles at the end of this book. The editors (Ed.) of this book are no authors. They have not modified or extended the original texts.

Pictures published in this book can be under different licences than the GNU Free Documentation License. You can get detailed informations about the authors and licences of pictures at the end of this book.

The content of this book was generated collaboratively by volunteers. Please be advised that nothing found here has necessarily been reviewed by people with the expertise required to provide you with complete, accurate or reliable information. Some information in this book maybe misleading or wrong. The Publisher does not guarantee the validity of the information found here. If you need specific advice (f.e. in fields of medical, legal, financial, or risk management questions) please contact a professional who is licensed or knowledgeable in that area.

Any brand names and product names mentioned in this book are subject to trademark, brand or patent protection and are trademarks or registered trademarks of their respective holders. The use of brand names, product names, common names, trade names, product descriptions etc. even without a particular marking in this works is in no way to be construed to mean that such names may be regarded as unrestricted in respect of trademark and brand protection legislation and could thus be used by anyone.

Cover image: www.ingimage.com
Concerning the licence of the cover image please contact ingimage.

Publisher:
Part Press is a trademark of
International Book Market Service Ltd., 17 Rue Meldrum, Beau Bassin, 1713-01 Mauritius
Email: info@bookmarketservice.com
Website: www.bookmarketservice.com

Published in 2012

Printed in: U.S.A., U.K., Germany. This book was not produced in Mauritius.

ISBN: 978-613-8-68886-0

Contents

Articles

References

Arkansas Highway 204

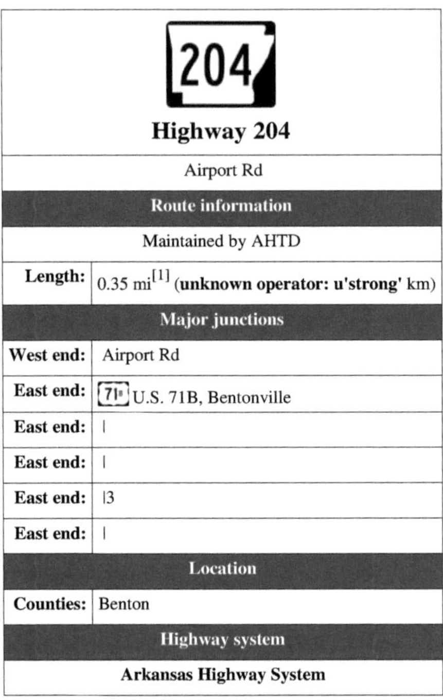

Highway 204

Airport Rd

Route information	
Maintained by AHTD	
Length:	0.35 mi[1] (**unknown operator: u'strong'** km)
Major junctions	
West end:	Airport Rd
East end:	71 U.S. 71B, Bentonville
East end:	I
East end:	I
East end:	I3
East end:	I
Location	
Counties:	Benton
Highway system	
Arkansas Highway System	

Arkansas Highway 204 (**AR 204** and **Hwy. 204**) is an east–west state highway in Benton County, Arkansas. The route of 0.35 miles (**unknown operator: u'strong'** km) runs from Bentonville Municipal Airport west to US 71B in Bentonville.[2]

Route description

The route begins at the Bentonville Municipal Airport and runs west to terminate at US 71B (Walton Blvd). The route is two-lane, undivided its entire length.[1] A 2010 study of average daily traffic determined that Highway 204 serves 3000 vehicles per day.[1] Bentonville Municipal Airport serves as a general use airport, supersceeded within Northwest Arkansas by Northwest Arkansas Regional Airport (XNA).

Major intersections

The entire route is in Bentonville, Benton County.

Mile[1]	Destinations	Notes
0.0	Bentonville Municipal Airport	western terminus
0.35	[71B] U.S. 71B (Walton Blvd)	eastern terminus
1.000 mi = 1.609 km; 1.000 km = 0.621 mi		

History

The highway was most recently paved in 1975.[1]

Former route

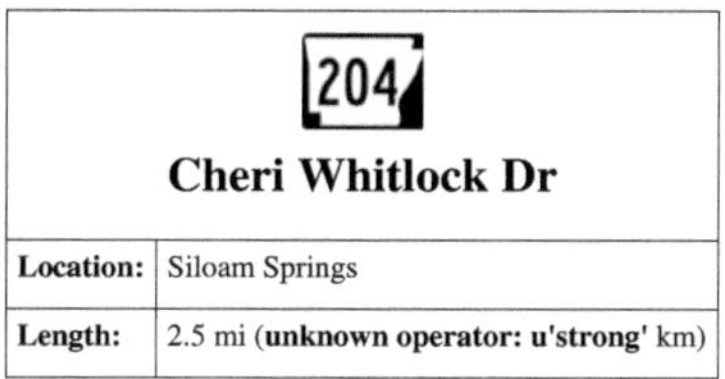

Cheri Whitlock Dr

Location:	Siloam Springs
Length:	2.5 mi (**unknown operator: u'strong'** km)

Arkansas Highway 204, known colloquially as **Cheri Whitlock Drive**[3] was a 2.5-mile (**unknown operator: u'strong'** km) long roadway just north of Siloam Springs. Its western terminus was AR 43 and its eastern Terminus was AR 59 approximately one mile north of U.S. Route 412. In the mid 1990s Highway 204 was redesignated as an extension of Highway 43.

References

[1] Planning and Research Division (2010). "Arkansas Road Log Database" (http://www.arkansashighways.com/planning_research/ technical_services/databases/Road Log Database.zip) (in English) (Database). Arkansas State Highway and Transportation Department. . Retrieved June 9, 2011.

[2] Arkansas State Highway and Transportation Department. *General Highway Map* (http://www.arkansashighways.com/maps/Counties/ County PDFs/BentonCounty.pdf) (Map) (Benton County ed.). . Retrieved June 12, 2011.

[3] Google. *Google Maps* (http://maps.google.com/maps?hl=en&tab=wl) (Map). . Retrieved April 13, 2009.

Bentonville Municipal Airport

<table>
<tr><td colspan="4" align="center">Bentonville Municipal Airport
Louise M. Thaden Field</td></tr>
<tr><td colspan="4" align="center">IATA: none – ICAO: KVBT – FAA LID: VBT</td></tr>
<tr><td colspan="4" align="center">Summary</td></tr>
<tr><td colspan="2">Airport type</td><td colspan="2">Public</td></tr>
<tr><td colspan="2">Owner</td><td colspan="2">City of Bentonville</td></tr>
<tr><td colspan="2">Serves</td><td colspan="2">Bentonville, Arkansas</td></tr>
<tr><td colspan="2">Hub for</td><td colspan="2">{{{hub}}}</td></tr>
<tr><td colspan="2">Elevation AMSL</td><td colspan="2">1,296 ft / 395 m</td></tr>
<tr><td colspan="2">Coordinates</td><td colspan="2">36°20′45″N 094°13′10″W</td></tr>
<tr><td colspan="4" align="center">Runways</td></tr>
<tr><td>Direction</td><td colspan="2" align="center">Length</td><td>Surface</td></tr>
<tr><td></td><td align="center">ft</td><td align="center">m</td><td></td></tr>
<tr><td align="center">18/36</td><td align="center">4,082</td><td align="center">1,244</td><td>Asphalt</td></tr>
<tr><td colspan="4" align="center">Statistics (2009)</td></tr>
<tr><td colspan="2">Aircraft operations</td><td colspan="2">18,100</td></tr>
<tr><td colspan="2">Based aircraft</td><td colspan="2">40</td></tr>
<tr><td colspan="4">Source: Federal Aviation Administration[1]</td></tr>
</table>

Bentonville Municipal Airport (ICAO: **KVBT**, FAA LID: **VBT**) is a city-owned, public-use airport located two nautical miles (3.7 km) south of the central business district of Bentonville, a city in Benton County, Arkansas, United States.[1] It is also known as **Louise M. Thaden Field**[1] or **Louise Thaden Field**, a name it was given in 1951 to honor Louise McPhetridge Thaden (1905–1979), an aviation pioneer from Bentonville.[2]

This airport is included in the FAA's National Plan of Integrated Airport Systems for 2009–2013, which categorizes it as a *general aviation* facility.[3] Although most U.S. airports use the same three-letter location identifier for the FAA and IATA, this airport is assigned **VBT** by the FAA but has no designation from the IATA.[4]

Facilities and aircraft

Bentonville Municipal Airport covers an area of 130 acres (**unknown operator: u'strong'** ha) at an elevation of 1,296 feet (395 m) above mean sea level. It has one runway designated 18/36 with an asphalt surface measuring 4,082 by 65 feet (1,244 x 20 m).[1]

For the 12-month period ending August 31, 2009, the airport had 18,100 aircraft operations, an average of 49 per day: 99% general aviation and 1% military. At that time there were 40 aircraft based at this airport: 85% single-engine, 12.5% multi-engine and 2.5% helicopter.[1]

References

[1] FAA Airport Master Record for VBT (http://www.gcr1.com/5010web/airport.cfm?Site=VBT) (Form 5010 (http://www.gcr1.com/5010web/Rpt_5010.asp?au=PU&o=PU&faasite=00886.8*A&fn=VBT) PDF). Federal Aviation Administration. Effective 11 February 2010.

[2] "Who Is Louise Thaden?" (http://www.summitaviationllc.com/airport_thaden.asp). Summit Aviation. . Retrieved 4 Mar 2010.

[3] National Plan of Integrated Airport Systems (http://www.faa.gov/airports/planning_capacity/npias/) for 2009–2013: Appendix A: Part 1 (PDF, 1.33 MB) (http://www.faa.gov/airports/planning_capacity/npias/reports/media/2009/npias_2009_appA_part1.pdf). Federal Aviation Administration. Updated 15 Oct 2008.

[4] "Bentonville Municipal Airport (IATA: none, ICAO: KVBT, FAA: VBT)" (http://www.gcmap.com/airport/KVBT). Great Circle Mapper. . Retrieved 4 Mar 2010.

External links

- Summit Aviation (http://www.summitaviationllc.com/), the fixed base operator (FBO)
- Aerial image as of 28 February 2001 (http://msrmaps.com/map.aspx?t=1&s=11&lat=36.3457&lon=-94.2193&w=600&h=800&lp=---+None+---) from USGS *The National Map*
- FAA Terminal Procedures for VBT (http://aeronav.faa.gov/digital_tpp_search.asp?fldIdent=VBT&fld_ident_type=FAA&ver=1204&eff=04-05-2012&end=05-03-2012&submit1=Search), effective 5 April 2012
- Resources for this airport:
    - AirNav airport information for KVBT (http://www.airnav.com/airport/KVBT)
    - FlightAware airport information (http://flightaware.com/resources/airport/KVBT) and live flight tracker (http://flightaware.com/live/airport/KVBT)
    - NOAA/NWS latest weather observations (http://www.crh.noaa.gov/data/obhistory/KVBT.html)
    - SkyVector aeronautical chart (http://skyvector.com/?id=KVBT&zoom=2), Terminal Procedures (http://skyvector.com/airport/VBT)

Benton County, Arkansas

Benton County, Arkansas	
Location in the state of Arkansas	
Arkansas's location in the U.S.	
Founded	30 September 1836
Seat	Bentonville
Largest city	Rogers
Area - Total - Land - Water	880.24 sq mi (2280 km²) 845.99 sq mi (2191 km²) 34.25 sq mi (89 km²), 3.89%
Population - **(2010)** - Density	221339 262/sq mi (101/km²)
Website	www.co.benton.ar.us [1]

Benton County is a county located in the U.S. state of Arkansas. As of the 2000 census, the population was 153,406. The U.S. Census Bureau 2010 population is 221,339.[2] The county seat is Bentonville.[3] Benton County was formed on 30 September 1836 and was named after Thomas Hart Benton, a U.S. Senator from Missouri. It is a dry county; alcohol sales are prohibited, except in establishments with a private club liquor license.

Benton County is part of the Fayetteville–Springdale–Rogers, AR-MO Metropolitan Statistical Area.

Geography

According to the 2000 census, the county has a total area of 880.24 square miles (**unknown operator: u'strong'** km^2), of which 845.99 square miles (**unknown operator: u'strong'** km^2) (or 96.11%) is land and 34.25 square miles (**unknown operator: u'strong'** km^2) (or 3.89%) is water.[4] Most of the water is in Beaver Lake.

Transportation

- Interstate 540
- U.S. Highway 62
- U.S. Highway 71
- U.S. Highway 412
- Highway 12

- ⟦16⟧ Highway 16
- ⟦43⟧ Highway 43
- ⟦59⟧ Highway 59
- ⟦72⟧ Highway 72
- ⟦94⟧ Highway 94
- ⟦102⟧ Highway 102

The Northwest Arkansas Regional Airport is located near Highfill.

The Arkansas and Missouri Railroad parallels US Highways 62 and 71 in the county.

The historic Trail of Tears is on US highways 62 and 71, connects with another US route 412 in nearby Washington County.

Adjacent counties

- Barry County, Missouri (north)
- Carroll County (east)
- Madison County (southeast)
- Washington County (south)
- Adair County, Oklahoma (southwest)
- Delaware County, Oklahoma (west)
- McDonald County, Missouri (northwest)

National protected areas

- Logan Cave National Wildlife Refuge
- Ozark National Forest (part)
- Pea Ridge National Military Park

Demographics

Historical populations		
Census	Pop.	%±
1840	2228	—
1850	3710	66.5%
1860	9306	150.8%
1870	13831	48.6%
1880	20328	47.0%
1890	27716	36.3%
1900	31611	14.1%
1910	33389	5.6%
1920	36253	8.6%
1930	35253	−2.8%
1940	36148	2.5%
1950	38076	5.3%
1960	36272	−4.7%

1970	50476	39.2%
1980	78115	54.8%
1990	97499	24.8%
2000	153406	57.3%
2010	221339	44.3%
[5] [6] [7]		

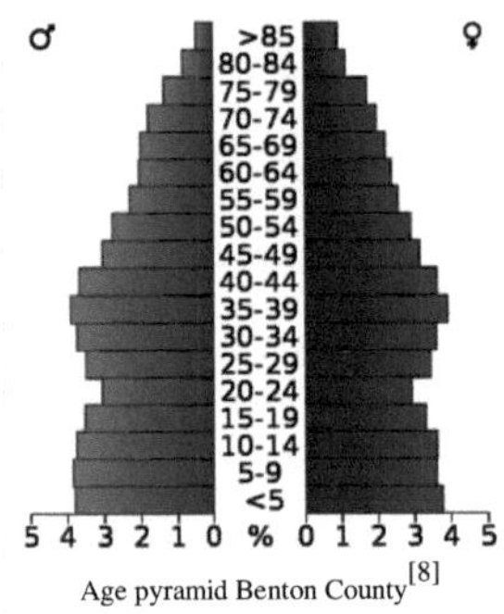

Age pyramid Benton County[8]

As of the census[9] of 2000, there were 153,406 people, 58,212 households, and 43,484 families residing in the county. The population density was 181 people per square mile (70/km²). There were 64,281 housing units at an average density of 76 per square mile (29/km²). The racial makeup of the county was 90.87% White, 0.41% Black or African American, 1.65% Native American, 1.09% Asian, 0.08% Pacific Islander, 4.08% from other races, and 1.82% from two or more races. 8.78% of the population were Hispanic or Latino of any race.

As of 2005 Benton County's population was 81.7% non-Hispanic white, while the percentage of Latinos grew by 60 percent in the time period. Latinos are attracted to the growth of light industrial jobs, home construction and service sector in the county. 1.1% of the population was African-American (perhaps the lowest in all of Arkansas); 1.6% was Native American (the historical presence of the Cherokee Indians live in close proximity to Oklahoma); 1.7% was Asian (there was a large influx of Filipinos, Vietnamese and South Asian immigrants arrived in recent decades) and 0.2% of the population was Pacific Islander. 1.6% reported two or more races, usually not black-white due to a minuscule African-American population. And 12.8% was Latino, but the United States Hispanic Chamber of Commerce believed the official estimate is underreported and Latinos could well be 20 percent of the population.[10]

There were 58,212 households out of which 34.40% had children under the age of 18 living with them, 63.00% were married couples living together, 8.20% had a female householder with no husband present, and 25.30% were non-families. 21.10% of all households were made up of individuals and 8.50% had someone living alone who was 65 years of age or older. The average household size was 2.60 and the average family size was 3.01.

In the county the population was spread out with 26.60% under the age of 18, 8.60% from 18 to 24, 29.40% from 25 to 44, 21.10% from 45 to 64, and 14.30% who were 65 years of age or older. The median age was 35 years. For every 100 females there were 97.40 males. For every 100 females age 18 and over, there were 94.90 males.

The median income for a household in the county was $40,281, and the median income for a family was $45,235. Males had a median income of $30,327 versus $22,469 for females. The per capita income for the county was $19,377. About 7.30% of families and 10.10% of the population were below the poverty line, including 13.80% of those under age 18 and 7.30% of those age 65 or over.

As of census 2010 the county population was 221,339. The racial makeup of the county was 76.18% Non-Hispanic white, 1.27% black, 1.69% Native American, 2.85% Asian, 0.30% Pacific Islander, 0.10% Non-Hispanics of some other race, 1.93% Non-Hispanics reporting two or more races and 15.49% Hispanic or Latino.

Benton County Corporations

- Wal-Mart corporate headquarters is located in Bentonville.
- Daisy Outdoor Products, known for its air rifles, is headquartered in Rogers.
- JB Hunt Transport Services corporate headquarters is located in Lowell.
- Tyson Foods, based in nearby Springdale, has a distribution center located in Rogers.

Cities and towns

- Avoca
- Bella Vista
- Bentonville
- Bethel Heights
- Cave Springs
- Centerton
- Decatur
- Elm Springs
- Garfield
- Gateway
- Gentry
- Gravette
- Highfill
- Little Flock
- Lowell
- Pea Ridge
- Rogers
- Siloam Springs
- Springdale (mostly in Washington County)
- Springtown
- Sulphur Springs
- War Eagle

Census Designated Places (CDPs)

- Hiwasse
- Lost Bridge Village

Townships

Note: Most Arkansas counties have names for their townships. Benton County, however, has numbers instead of names.

Townships in Arkansas are the divisions of a county. Each township includes unincorporated areas and some may have incorporated towns or cities within part of their space. Townships have limited purposes in modern times. However, the US Census does list Arkansas population based on townships (often referred to as "minor civil divisions"). Townships are also of value for historical purposes in terms of genealogical research. Each town or city is within one or more townships in an Arkansas county based on census maps. The townships

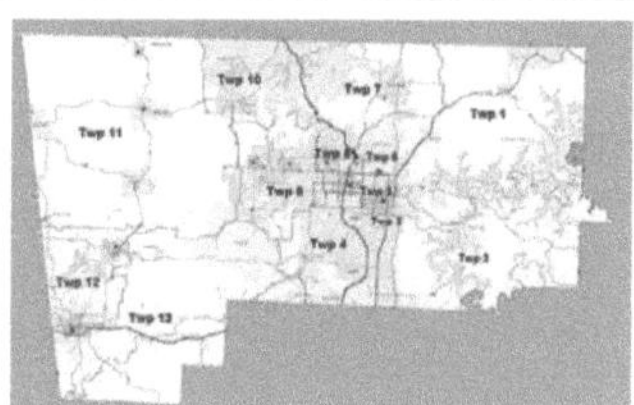

Townships in Benton County, Arkansas as of 2010

of Benton County are listed below with the town(s) and/or city that are fully or partially inside them listed in parentheses. [11] [12]

- **Twp 1** (all of the following: Garfield, Gateway, Lost Bridge Village, Prairie Creek; parts of the following: Avoca, Rogers)
- **Twp 2** (small parts of the following: Lowell, Rogers, Springdale)
- **Twp 3** (parts of the following: Lowell, Rogers, Springdale; most of Bethel Heights)
- **Twp 4** (all of Cave Springs ; most of the following: Lowell, Rogers, Springdale (within Benton County); small parts of Elm Springs)
- **Twp 5** (part of Rogers)
- **Twp 6** (most of Little Flock; almost half of Avoca; small parts of Bentonville, Pea Ridge, Rogers)
- **Twp 7** (most of Pea Ridge; part of Bella Vista; small part of Bentonville)
- **Twp 8** (part of Bentonville)
- **Twp 9** (most of the following: Bentonville, Centerton; small part of Highfill)
- **Twp 10** (most of the following: Bella Vista, Hiwasse)

- **Twp 11** (all of the following: Cherokee City, Decatur, Gravette, Maysville, Sulphur Springs; small parts of the following: Centerton, Highfill, Hiwasse)
- **Twp 12** (most of Gentry; more than half of Siloam Springs;
- **Twp 13** (all of Springtown; most of Highfill; small parts of the following: Elm Springs, Gentry, Springdale)

See also

- National Register of Historic Places listings in Benton County, Arkansas

References

[1] http://www.co.benton.ar.us

[2] Benton County, Arkansas - Population Finder - American FactFinder (http://factfinder.census.gov/servlet/
     SAFFPopulation?_event=Search&_name=benton+county&_state=04000US05&_county=benton+county&_cityTown=benton+county&
     _zip=&_sse=on&_lang=en&pctxt=fph)

[3] "Find a County" (http://www.naco.org/Counties/Pages/FindACounty.aspx). National Association of Counties. . Retrieved 2011-06-07.

[4] "Census 2000 U.S. Gazetteer Files: Counties" (http://www.census.gov/tiger/tms/gazetteer/county2k.txt). United States Census. .
     Retrieved 2011-02-13.

[5] http://www.census.gov/population/www/censusdata/cencounts/files/ar190090.txt

[6] http://factfinder2.census.gov

[7] http://mapserver.lib.virginia.edu/

[8] Based on 2000 census data

[9] "American FactFinder" (http://factfinder.census.gov). United States Census Bureau. . Retrieved 2008-01-31.

[10] Benton County QuickFacts from the US Census Bureau (http://quickfacts.census.gov/qfd/states/05/05007.html)

[11] US Census Bureau. *2011 Boundary and Annexation Survey (BAS): Benton County, AR* (http://www2.census.gov/geo/pvs/bas/bas11/
     st05_ar/cou/c05007_benton/BAS11C20500700000_000.pdf) (Map). . Retrieved 20110808.

[12] http://www.census.gov/geo/www/maps/DC10_GUBlkMap/cousub/dc10blk_st05_cousub.html#B

External links

- Official Website of Benton County, Arkansas (http://www.co.benton.ar.us/)
- Benton County Code of Ordinances (http://www.municode.com/resources/gateway.asp?pid=12635&sid=4/)
- County Records Online (http://www.arcountydata.com)

Siloam Springs, Arkansas

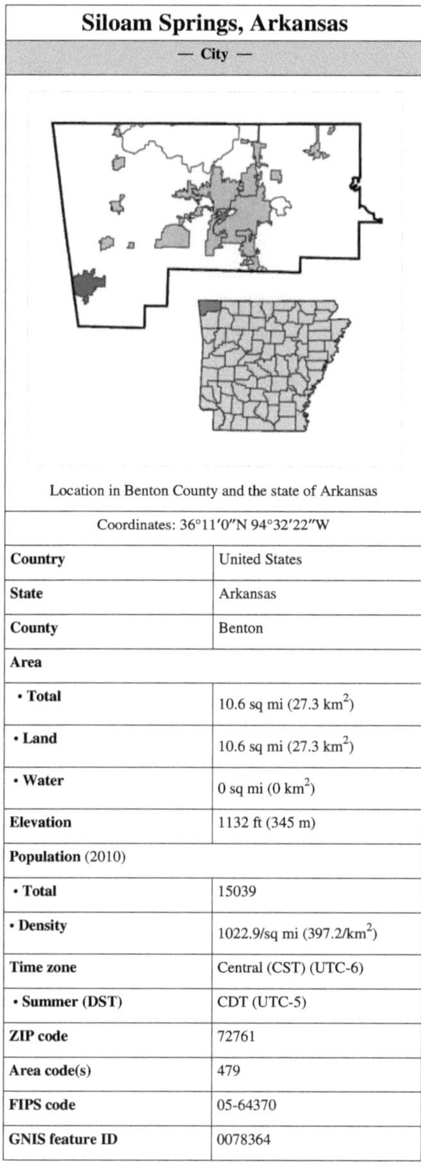

Siloam Springs, Arkansas	
— City —	
Location in Benton County and the state of Arkansas	
Coordinates: 36°11′0″N 94°32′22″W	
Country	United States
State	Arkansas
County	Benton
Area	
• Total	10.6 sq mi (27.3 km^2)
• Land	10.6 sq mi (27.3 km^2)
• Water	0 sq mi (0 km^2)
Elevation	1132 ft (345 m)
Population (2010)	
• Total	15039
• Density	1022.9/sq mi (397.2/km^2)
Time zone	Central (CST) (UTC-6)
• Summer (DST)	CDT (UTC-5)
ZIP code	72761
Area code(s)	479
FIPS code	05-64370
GNIS feature ID	0078364

Siloam Springs is a city in Benton County, Arkansas, United States. According to 2010 Census Bureau estimates, the population of the city is 15,039.[1] It is part of the Fayetteville–Springdale–Rogers, AR-MO Metropolitan Statistical Area.

Siloam Springs is home to John Brown University, a private, interdenominational, Christian liberal arts college. In September 2011, Siloam Springs became a Main Street Community, as recognized by the Arkansas Historic Preservation Program for efforts to preserve and revitalize the city's historic district.

Geography

Siloam Springs is located at 36°11′0″N 94°32′22″W (36.183359, -94.539315)[2].

According to the United States Census Bureau, the city has a total area of 10.6 square miles (**unknown operator: u'strong'** km^2), all land.

The area is located in the region of the country where the southern plains meet the Ozark Mountains. The city sits atop a plateau. Numerous dogwood trees grow across the landscape. The area is a mix of southern and midwestern cultures. Osage Indians were the first inhabitants of the area. Siloam Springs' first settlers were of German and Scots-Irish origin. Simon Sager is considered the initial founder of the first settlement known as Hico. A small creek, named after the founder, Sager Creek, flows through the downtown area. Siloam Springs is made up of Siloam Springs, Arkansas, and West Siloam Springs, Oklahoma. The latter is in the territory of the Cherokee Nation in northeastern Oklahoma.

Demographics

Downtown Fountains

As of the census[3] of 2010, there were 15,039 people in 5,138 households with 93.3% of the population in households. The racial and ethnic composition of the population was 68.6% non-Hispanic white, 0.8% black, 4.6% Native American, 1.6% Asian, 0.2% non-Hispanic reporting some other race, 5.0% from two or more races and 20.8% Hispanic or Latino.[4]

At the 2000 census there were 2,647 families residing in the city. The population density was 1,027.2 per square mile (396.4/km²). There were 4,223 housing units at an average density of 400.1 per square mile (154.4/km²). The racial makeup of the city was 85.22% White, 0.49% Black or African American, 4.29% Native American, 0.83% Asian, 0.08% Pacific Islander, 5.67% from other races, and 3.42% from two or more races. 14.00% of the population were Hispanic or Latino of any race.

There were 3,894 households out of which 34.8% had children under the age of 18 living with them, 53.8% were married couples living together, 10.6% had a female householder with no husband present, and 32.0% were non-families. 26.9% of all households were made up of individuals and 11.9% had someone living alone who was 65 years of age or older. The average household size was 2.57 and the average family size was 3.11.

In the city the population was spread out with 26.0% under the age of 18, 16.8% from 18 to 24, 27.8% from 25 to 44, 17.1% from 45 to 64, and 12.3% who were 65 years of age or older. The median age was 30 years. For every 100 females there were 95.6 males. For every 100 females age 18 and over, there were 91.2 males.

The median income for a household in the city was $34,513, and the median income for a family was $41,153. Males had a median income of $27,339 versus $21,451 for females. The per capita income for the city was $16,047. About 9.5% of families and 12.5% of the population were below the poverty line, including 17.6% of those under age 18 and 8.6% of those age 65 or over.

As of 2009, there were 52 churches that called Siloam Springs home by address. There are reports that Siloam Springs has a record for most number churches per capita, and while the ratio is higher than average, it has never been verified through reliable documentation. Major employers in Siloam Springs include Gates Corporation, La-Z-Boy, DaySpring (a subsidiary of Hallmark Cards), Allen Canning, Cobb-Vantress, and John Brown University.

Notable people

Siloam Springs is the birthplace and former residence of award-winning contemporary Cherokee basketry artist Mike Dart.

Preston Bynum, a lobbyist in Little Rock, served as the state representative from Siloam Springs from 1969 to 1980. He was the second Republican to represent Benton County in the legislature in the 20th century, the first having been Jim Sheets.

Alice Ghostley, an actress, spent a number of years in Siloam Springs as a youth.

Government

Siloam Springs has a City Administrator form of government. The government body consists of the Mayor, Board of Directors and District [[Judge]. All positions are chosen by election. The other officials and commissioners are appointed with Board approval.

Siloam Springs government positions[5]

Name	Position	Seat
David Allen	Mayor	N/A
John Turner	City Board of Directors	Ward 1
James Fuller	City Board of Directors	Ward 2
Ken Krafft	City Board of Directors	Ward 3
Judy Nation	City Board of Directors	Ward 4
Ken Wiles	City Board of Directors	Position 5
Carol Smiley	City Board of Directors	Position 6
Mark Long	City Board of Directors	Position 7

Tourist events

Event	Time of Year	Attendance (approx.)
Dogwood Festival	April - Last Weekend	30,000
Siloam Springs Rodeo	June	10,000
Christmas Parade	December (1st Saturday)	6,500
City Fireworks Presentation	July 4	6,000
Northwest Arkansas Marching Band Invitational	October	2,500
JBU Candlelight Christmas Concerts	December	3,000
JBU Homecoming	October	1,000
Siloam Springs Music Games (Marching Band Competition)	July	2,000
Sager Creek Arts Center	All year	10,000 per year

References

[1] "Annual Estimates of the Population for All Incorporated Places in Arkansas" (http://quickfacts.census.gov/qfd/states/05/0564370.html) (CSV). *2005 Population Estimates*. U.S. Census Bureau, Population Division. December 9,2011. . Retrieved November 16, 2006.

[2] "US Gazetteer files: 2010, 2000, and 1990" (http://www.census.gov/geo/www/gazetteer/gazette.html). United States Census Bureau. 2011-02-12. . Retrieved 2011-04-23.

[3] "American FactFinder" (http://factfinder.census.gov). United States Census Bureau. . Retrieved 2008-01-31.

[4] 2010 general profile of population and housing characteristics for Siloam Springs from the US census

[5] http://www.siloamsprings.com/government/board/

External links

- Official site (http://www.siloamsprings.com/)
- History of Siloam Springs (http://www.siloamsprings.com/about/history.php)
- Encyclopedia of Arkansas History & Culture entry: Siloam Springs (Benton County) (http://www. encyclopediaofarkansas.net/encyclopedia/entry-detail.aspx?entryID=838)

Northwest Arkansas Regional Airport

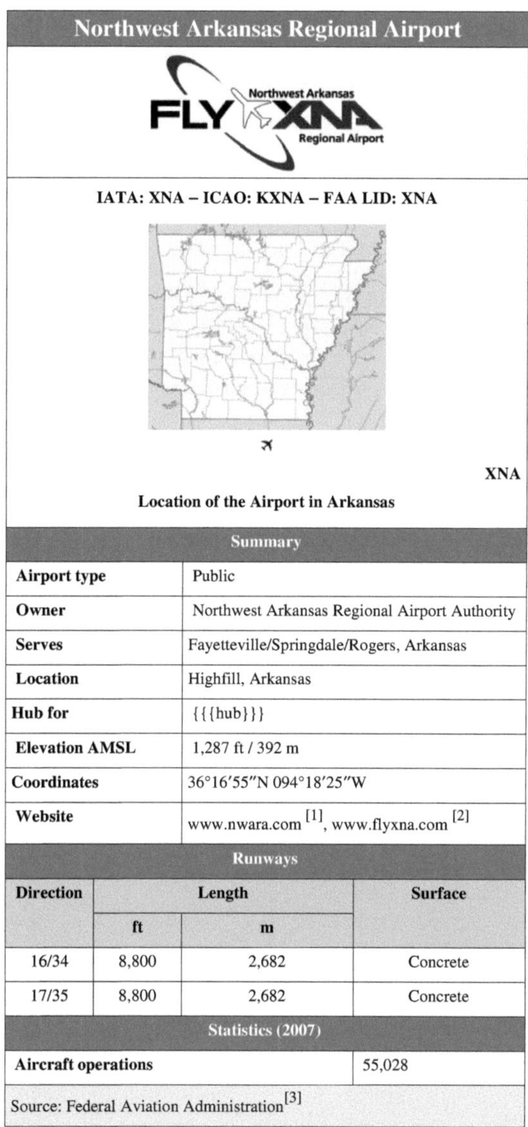

Northwest Arkansas Regional Airport	
IATA: XNA – ICAO: KXNA – FAA LID: XNA	
Location of the Airport in Arkansas	
Summary	
Airport type	Public
Owner	Northwest Arkansas Regional Airport Authority
Serves	Fayetteville/Springdale/Rogers, Arkansas
Location	Highfill, Arkansas
Hub for	{{{hub}}}
Elevation AMSL	1,287 ft / 392 m
Coordinates	36°16′55″N 094°18′25″W
Website	www.nwara.com [1], www.flyxna.com [2]

Runways

Direction	Length		Surface
	ft	m	
16/34	8,800	2,682	Concrete
17/35	8,800	2,682	Concrete

Statistics (2007)

Aircraft operations	55,028

Source: Federal Aviation Administration[3]

Northwest Arkansas Regional Airport (IATA: **XNA**, ICAO: **KXNA**, FAA LID: **XNA**) is an airport located in Highfill, Arkansas, near Bentonville, Rogers, Fayetteville, Springdale, and Siloam Springs, Arkansas. It is commonly referred to by its IATA code, which is incorporated in the airport's logo as "Fly XNA".

XNA opened in November 1998 as a replacement airport for commercial traffic previously served by Fayetteville's aging and inadequate Drake Field, which was undersized to serve the rapidly-growing Northwest Arkansas region. American Airlines is the major carrier, formerly serving 6 cities, and currently 3 from XNA. Much of AA's expansion at XNA is due to its contract with Wal-Mart, which is based in Bentonville.

Expansion

Due to rapid growth in Northwest Arkansas, in 2007 airport officials announced the construction of a new concourse costing between $20–$25 million. The new concourse is east of the upper concourse, allowing the airport to park eight additional planes for boarding. The airport previously had 12 airplane parking positions. It took over three years to complete.[4]

The airport recently completed a ticket counter expansion, and is planning two large warehouse additions. The airport has completed a $21 million expansion to the upper-level concourse that includes the state's first moving walkway. With the walkway it will only take about three minutes to get from security to the last gate. The addition adds 51000 square feet (**unknown operator: u'strong'** m^2) and 8 upper level gates to the east side of the airport.[5]

Facilities and aircraft

Northwest Arkansas Regional Airport covers an area of **unknown operator: u','** acres (**unknown operator: u'strong'unknown operator: u','**ha) at an elevation of 1,287 feet (392 m) above mean sea level. It has two runways designated 16/34 and 17/35 with a concrete surface measuring 8,800 by 150 feet (2,682 x 46 m). For the 12-month period ending August 30, 2007, the airport had 55,028 aircraft operations, an average of 150 per day: 46% air taxi, 26% military, 15% general aviation and 13% scheduled commercial.[3]

Statistics

Statistics for Northwest Arkansas Regional Airport

Year	Total Passengers (enplane and deplane)	Total Cargo (tons)	Total operations
1998	53,565*	5,618*	5,238*
1999	653,022	205,175	38,548
2000	725,175	337,143	40,169
2001	735,822	356,186	40,426
2002	786,948	371,987	37,282
2003	892,489	329,304	44,259
2004	1,020,146	348,171	49,379
2005	1,168,858	296,740	55,091
2006	1,172,049	164,121	53,522
2007	1,200,122	230,214	55,193
2008	1,146,954	253,411	47,975
2009	1,084,471	196,706	44,568
2010	1,139,801	96,703	44,960
2011+	766,173	38,891	28,205

- (*) Note: Statistics for 1998 include only November and December.[6]

- (+) Note: First eight months.

Airlines

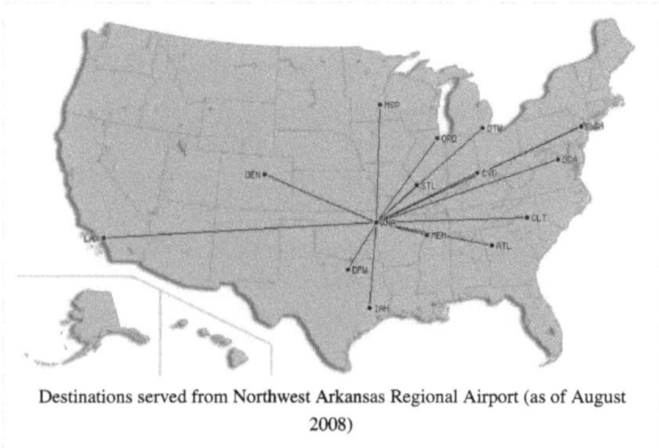

Destinations served from Northwest Arkansas Regional Airport (as of August 2008)

Airlines	Destinations
Allegiant Air	Las Vegas, Los Angeles, Orlando-Sanford, Phoenix/Mesa
American Airlines	Dallas/Fort Worth
American Eagle	Chicago-O'Hare, Dallas/Fort Worth, New York-LaGuardia
Delta Connection operated by Comair	Atlanta, Cincinnati, Detroit
Delta Connection operated by ExpressJet	Atlanta, Memphis
Delta Connection operated by Pinnacle Airlines	Atlanta, Cincinnati, Detroit, Minneapolis/St. Paul
United Express operated by ExpressJet Airlines	Chicago-O'Hare, Denver, Houston-Intercontinental, Newark
United Express operated by SkyWest Airlines	Chicago-O'Hare, Denver
US Airways Express operated by PSA Airlines	Charlotte, Washington-National [begins July 11, 2012]

See also

- Fayetteville-Springdale-Rogers metropolitan area

References

[1] http://www.nwara.com/

[2] http://www.flyxna.com/

[3] FAA Airport Master Record for XNA (http://www.gcr1.com/5010web/airport.cfm?Site=XNA) (Form 5010 (http://www.gcr1.com/5010web/Rpt_5010.asp?au=PU&o=PU&faasite=00975.01*A&fn=XNA) PDF), effective 2008-09-25.

[4] "XNA Opens New Concourse" (http://nwahomepage.com/fulltext-news/?nxd_id=264017). *KNWA*. August 24, 2011.

[5] "XNA Expands Terminal, Runway Facilities" (http://www.4029tv.com/news/24103389/detail.html). *4029tv.com*. June 30, 2010.

[6] "Enplanements" (http://www.nwara.com/files/enplanements.htm). *Official website of Northwest Arkansas Regional Airport*.

External links

- Northwest Arkansas Regional Airport (http://www.nwara.com/), official site
- FAA Airport Diagram (http://aeronav.faa.gov/d-tpp/1204/09274AD.PDF) (PDF), effective 5 April 2012
- FAA Terminal Procedures for XNA (http://aeronav.faa.gov/digital_tpp_search.asp?fldIdent=XNA& fld_ident_type=FAA&ver=1204&eff=04-05-2012&end=05-03-2012&submit1=Search), effective 5 April 2012
- Resources for this airport:
  - AirNav airport information for KXNA (http://www.airnav.com/airport/KXNA)
  - ASN accident history for XNA (http://aviation-safety.net/database/airport/airport.php?id=XNA)
  - FlightAware airport information (http://flightaware.com/resources/airport/KXNA) and live flight tracker (http://flightaware.com/live/airport/KXNA)
  - NOAA/NWS latest weather observations (http://www.crh.noaa.gov/data/obhistory/KXNA.html)
  - SkyVector aeronautical chart for KXNA (http://skyvector.com/perl/code?id=KXNA&scale=2)
  - FAA current XNA delay information (http://www.fly.faa.gov/flyfaa/flyfaaindex.jsp?ARPT=XNA&p=0)

Bannered routes of U.S. Route 71

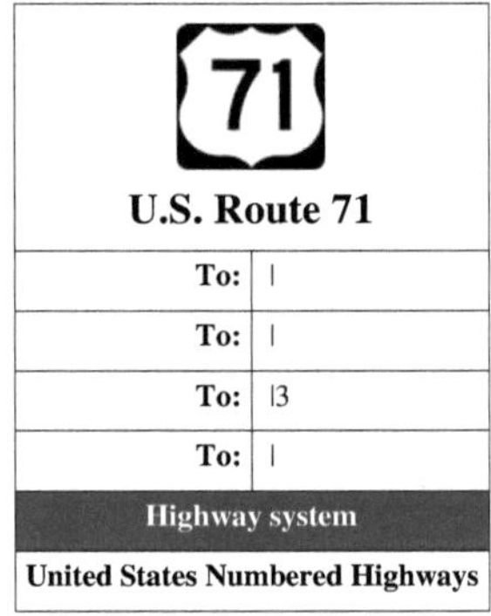

U.S. Route 71

To:	I
To:	I
To:	I3
To:	I
Highway system	
United States Numbered Highways	

A total of fourteen **bannered routes of U.S. Route 71** exist.

Alexandria bypass

U.S. Highway 71 Bypass

Location:	Alexandria, Louisiana

Bypass 71 is a controlled access highway at Alexandria, Louisiana. Its northern terminus is an interchange with Interstate 49 and U.S. Routes 71 and 165 north of Alexandria. Its southern terminus is at an interchange with Interstate 49 and U.S. Routes 71 and 167 south of Alexandria. Bypass 71 runs a total distance of approximately 6 miles (**unknown operator: u'strong'** km) and is concurrent with I-49 its whole length. It is signed at both the north

and south termini as By-Pass US 71.

Alexandria business loop

**U.S. Highway 71
Business**

Location:	Alexandria, Louisiana

Waldron business loop

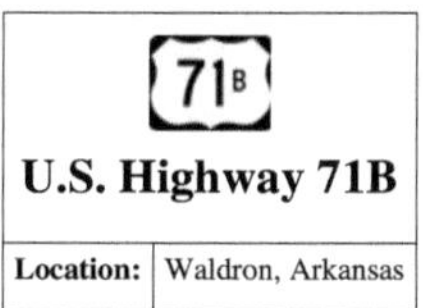

U.S. Highway 71B

Location:	Waldron, Arkansas

Business U.S. 71 in Waldron runs approximately 7 miles (**unknown operator: u'strong'** km) beginning at U.S. 71 2½ miles north of Waldron and ending at U.S. 71 4½ miles south of Waldron. Signed locally as Main Street, it was created in 1971 after U.S. 71 was rerouted around the west side of town.

Fort Smith business loop

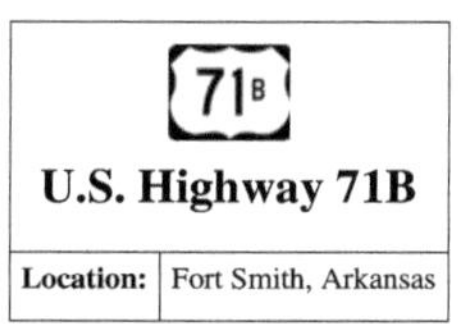

U.S. Highway 71B

Location:	Fort Smith, Arkansas

U.S. 71 Business runs approximately 13 miles (**unknown operator: u'strong'** km) between Alma, Arkansas and Fort Smith, Arkansas Its northern terminus is at Interstate 40 and U.S. Highway 71 at Alma and its southern terminus is at Interstate 540 and U.S. 71 in south Fort Smith. Highway 71 Business passes through the towns of Alma, Van Buren and Fort Smith. One half mile south of I-40 at Alma, U.S. 71 Business intersects U.S. Highway 64 and overlaps it to downtown Fort Smith and the junction of Arkansas Highway 22.

Within the city of Fort Smith, U.S. 71 Business is commonly referred to by the names of Midland Boulevard (north of downtown), North 10th (one-way north-to-south) & North 11th Streets (one-way south-to-north) within downtown, Towson Avenue (south of downtown) and Zero Street (beginning at the intersection of U.S. Highway 271 and Arkansas Highway 255).

Northwest Arkansas business route

U.S. Highway 71B	
Location:	Northwest Arkansas
Length:	34.27 mi[1] (**unknown operator: u'strong'** km)
Existed:	1971–present

U.S. Route 71 Business in Fayetteville–Springdale–Rogers Metropolitan Area is a business route of U.S. Route 71 that spans 34.27 miles (**unknown operator: u'strong'** km).[2] [3] US 71B begins in south Fayetteville when US 71 angles west to become the Fulbright Expressway with US 62 and I-540 (Future I-49).[4] US 71B meets AR 265 (Cato Springs Rd.), AR 16 and AR 180 before entering downtown Fayetteville. US 71B becomes a main promenade through town as College Avenue passing AR 45 along the way

As College Avenue, US 71B passes the Evelyn Hills Shopping Center, Fiesta Square, Spring Creek Centre, the Northwest Arkansas Mall, and Lake Fayetteville. Entering Springdale the route crosses by the Springdale Country Club before meeting and briefly concurring with US 412. After this, US 71B continues north to meet AR 264 in Bethel Heights.

US 71B becomes Bloomington Street in Lowell as it continues north to New Hope Road, AR 94. Entering Rogers US 71B meets US 62/AR 12 becoming Walnut Street and turning east. Walnut Street continues by St. Mary's Hospital and Dixieland Mall to cross over I-540 to enter Bentonville.

US 71B meets AR 112, AR 72, and AR 102 near the Bentonville Municipal Airport when arrowing north. Becoming Walton Boulevard, US 71B crosses Central Avenue before again meeting I-540/US 71 and terminating at Exit 93.

13.28

Major intersections

Southern terminus of US 71B in south Fayetteville.

US 71B intersects Martin Luther King Jr Blvd in Fayetteville at a busy intersection.

US 71B forms a brief concurrency with US 412 in Springdale.

County	Location	Mile[2] [3] [1]	Destinations	Notes
Washington	Fayetteville	0.0	71 U.S. 71	southern terminus
		1.45	16 Hwy. 16 (W 15th St)	
		2.04	180 Hwy. 180 west (Martin Luther King Jr. Blvd)	
		2.98	Dickson St.	
		3.14	45 Hwy. 45 east (E Lafayette St)	
		5.00	Township St.	former Highway 180
		6.75	To 540 62 71 I-540 / U.S. 62 / U.S. 71 (Fulbright Expy)	also access to Gregg Avenue
		7.0	Joyce Blvd 540 71 62 To I-540 / U.S. 71 / U.S. 62 (Fulbright Expy)	northbound has jughandle entrance to Fulbright Expy
	Springdale	10.07	412 To U.S. 412 east (W Robinson Ave)	connector road
		10.28	412 U.S. 412 west (S Thompson Ave)	US 412 concurrency begins
		10.80	412 U.S. 412 west (Sunset Ave)	US 412 concurrency ends

Benton	Bethel Heights	14.30	 Hwy. 264 east (Jackson Ave)	
	Lowell	16.29	Hwy. 264 west (W Monroe Ave)	
	Rogers		Hwy. 94 south (W New Hope Rd)	
			Hwy. 12 / Hwy. 94 north (N 8th St)	AR 12 concurrency begins, AR 94 concurrency ends
	Bentonville		I-540 / U.S. 71 / U.S. 62 (Fulbright Expy)	
			Hwy. 112 north (SE J St)	AR 112 concurrency south
			Hwy. 12 north / Hwy. 112 (SW Regional Airport Blvd)	AR 12/AR 112 concurrency ends
			Hwy. 204 west (Airport Rd)	to Bentonville Municipal Airport
			Hwy. 102 (SW 14th St)	
			Hwy. 72 west (SW 2nd St)	
			I-540 south / U.S. 71 (Bella Vista Way)	northern terminus
		1.000 mi = 1.609 km; 1.000 km = 0.621 mi		

Pineville–Anderson business loop

BUSINESS

(71)

U.S. Route 71 Business

| Location: | Pineville–Anderson |

U.S. Route 71 Business is an alternate alignment of U.S. Route 71 in southwest Missouri. Its northern terminus is at a partial interchange with US 71 approximately 7 miles (**unknown operator: u'strong'** km) north of Anderson. Its southern terminus is an at-grade intersection with US 71 and Wolf Den Road (also known as McDonald County Road 71-22B SW) approximately 2 miles (**unknown operator: u'strong'** km) south of Pineville. In Anderson, Business 71 runs concurrently with Route 59 for approximately 6 miles (**unknown operator: u'strong'** km) and Route 76 for 2 miles (**unknown operator: u'strong'** km). Business 71 was originally created in 2005 running from 7 miles (**unknown operator: u'strong'** km) north of Anderson to 2 miles (**unknown operator: u'strong'** km) south of town at an at-grade intersection with US 71. In 2007, it was extended along the former US 71 in Pineville after a new freeway section was built bypassing the town.

Neosho business loop

BUSINESS

(71)

U.S. Route 71 Business

Location:	Neosho, Missouri

U.S. Route 71 Business is a business route of U.S. Route 71. It begins two miles (3 km) northwest of Neosho, Missouri at US 71 and Route 86 and ends south of the city at US 71 and Route AA. Part of the highway overlaps Highway 86, another part overlaps Route 59. The road is a previous alignment of US 71, which has been realigned west of Neosho as a freeway.

Route 175 which ends near the north end of Business US 71 was once also labeled Business US 71, joining this route with Business US 71 in Joplin making this an even longer route.

Joplin business loop

**U.S. Route 71
Business**

Location:	Joplin, Missouri

Business US Highway 71 is an alternate alignment of U.S. Route 71 which begins at a junction of U.S. Route 71, Route 96, Route 171, and Route 571 in Carthage, Missouri. Its southern end is at the junction of US 71 and Route 175 about midway between Joplin and Neosho at Tipton Ford.

This particular section of Business US 71 is a relatively long one, and begins its existence as a freeway, finally becoming a non-freeway divided highway at Webb City. It is a major boulevard in Joplin where it is known as Rangeline Road, then continues south of Interstate 44 until it ends at US 71. The road was an earlier alignment of US 71 until the newer freeway was built to carry the road to the east of Joplin. At one point, Route 175 had been labeled Business US 71 and the road continued on as Business 71 in Neosho.

Cities on Business US 71 (Joplin):

- Carthage, Missouri
- Carterville, Missouri
- Webb City, Missouri
- Joplin, Missouri
- Silver Creek, Missouri
- Leawood, Missouri
- Saginaw, Missouri
- Tipton Ford, Missouri

Nevada business loop

**U.S. Route 71
Business**

Location:	Nevada, Missouri

Business U.S. Route 71 is a route in Nevada, Missouri, in the USA. It begins at U.S. Route 71 on the north side of the city and ends at the same highway in the east-central part of town.

Business 71 is a former alignment of US 71 through town. It begins as a set of exit/entrance ramps before becoming Osage Boulevard. It meets US 54 at Austin Blvd. and turns east. US 54 leaves the concurrency at Centennial Boulevard. while Business 71 continues west on Austin until meeting again with US 71.

This is one of the shortest alignments of the many US 71 business routes throughout western Missouri.

St. Joseph business loop

Maryville business loop

Business 71 is a former alignment of US 71 through Maryville, Missouri. Running a distance of approximately 5 miles (**unknown operator: u'strong'** km), its southern terminus is at US 71 south of Maryville. Its northern terminus is an intersection with US 71 and US 136 north of Maryville.

Clarinda business loop

U.S. Route 71 Business is a former alignment of US 71 through Clarinda, Iowa. It begins at the junction of U.S. Route 71 and Iowa Highway 2 in southern Clarinda. Then, it follows 16th Street (Glenn Miller Avenue) towards downtown Clarinda. At Washington Street, US 71 Business meets Iowa 2 Business, and both routes continue east, eventually leaving Clarinda. East of Clarinda, US 71 Bus./IA 2 Bus. intersect US 71. US 71 Business ends while Iowa 2 Business continues south to complete its business loop.

The entire route is in Clarinda, Page County.

Mile	Destinations	Notes
	[71][2] US 71 / Iowa 2	
	[2] Iowa 2 Bus. west	South end of IA 2 Bus. overlap
	[71][2] US 71 / Iowa 2 Bus. east	North end of IA 2 Bus. overlap
1.000 mi = 1.609 km; 1.000 km = 0.621 mi		
Concurrency terminus • Closed/former • Incomplete access • Unopened		

Storm Lake business loop

U.S. Route 71 Business is a former alignment of US 71 in Storm Lake, Iowa. It begins at the junction of U.S. Route 71 and Iowa Highway 7. It follows Iowa 7 into Storm Lake along Lakeshore Drive, Flindt Drive, and Milwaukee Avenue. At Lake Street, US 71 Business turns north, leaving Iowa 7, where it eventually leaves Storm Lake. Near Truesdale, US 71 Bus. turns east and rejoins US 71.

The entire route is in Buena Vista County.

Location	Mile	Destinations	Notes
Storm Lake		[71][7] US 71 / Iowa 7 west	South end of IA 7 overlap
		[7] Iowa 7 west (Milwaukee Avenue)	North end of IA 7 overlap
Truesdale		[71] US 71	
1.000 mi = 1.609 km; 1.000 km = 0.621 mi			
Concurrency terminus • Closed/former • Incomplete access • Unopened			

Willmar business loop

Former routes

Joplin alternate

**U.S. Route 71
Alternate**

Location:	Joplin, Missouri

Former Alternate US 71 near Carthage, Missouri.
This section is now a frontage road ("outer road"
in Missouri) of US 71.

Alternate US 71 was a former bannered highway which provided an alternate route for US 71 between Carthage, Missouri and Neosho, Missouri, bypassing Joplin, Missouri. Both endpoints were junctions with US 71. This section of road from Fidelity to Carthage was originally Route 38, renumbered **Route 38N** in about 1930.

At Carthage, Alternate 71 followed what is currently Route 571 to US 71 at what is now the intersection of Route 96/571. When the freeway was built around Carthage, it ended at that the current exit of 71 at Route 96/571 & Business 71.

In 1999, the Alternate 71 designation was deleted. The section north of Interstate 44 at Fidelity was redesignated US 71, with the former US 71 being designated Business 71. South of Interstate 44, it continues as Route 59 to U.S. Route 60. From there it followed US 60 to Neosho. Other than its endpoints, only two towns were located on the former highway: Fidelity, Missouri and Diamond, Missouri.

The road was originally assigned at Optional US 71 in 1932 and changed to Alternate US 71 in 1935.

Kansas City bypass

**U.S. Route 71
Bypass**

Location:	Kansas City, Missouri
Existed:	1932–1969

U.S. Route 71 Bypass (Kansas City) was the original name for a highway that connected Harrisonville, Missouri to just south of Platte City, Missouri, where it rejoined US 71 Highway near Kansas City International Airport. When I-29 was opened in the mid-1960s, it was renumbered Route 291.

At Lee's Summit, Missouri, it connects to I-470. It remains concurrent with the Interstate, until I-470 terminates at I-70. Route 291 continues to the north after its junction with I-70. The route has been rerouted several times, and has seen improvements over the years, and continues to be a major highway in eastern Jackson County, Missouri. In Platte and Clay Counties, it also is known as Cookingham Drive, and Mid Continent Trafficway.

Platte City bypass

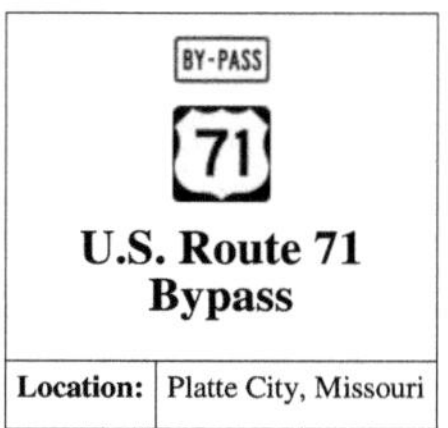

**U.S. Route 71
Bypass**

Location:	Platte City, Missouri

References

[1] "[Arkansas] State Highways 2009 (Database)." April 2010. AHTD: Planning and Research Division. |length_notes=mileage includes flyover
    of 1.83 miles (**unknown operator: u'strong'** km) serving I-540/US 71/US 62 Database. (http://www.arkansashighways.com/
    planning_research/technical_services/databases/Arkansas_Roadlog_2009.zip) Retrieved June 5, 2011.

[2] Arkansas State Highway and Transportation Department. AHTD Washington County map (http://www.arkansashighways.com/maps/
    counties/county PDFs/WashingtonCounty.pdf#) Retrieved on June 24, 2009.

[3] Arkansas State Highway and Transportation Department. AHTD Benton County map (http://www.arkansashighways.com/maps/counties/
    county PDFs/BentonCounty.pdf#) Retrieved on June 24, 2009.

[4] "Interstate 49." Profile of I-49. (http://www.interstate-guide.com/i-049.html) Retrieved on June 24, 2009.

Arkansas State Highway and Transportation Department

<table>
<tr><td colspan="2" align="center">Arkansas State Highway and Transportation
Department (AHTD)</td></tr>
<tr><td colspan="2" align="center">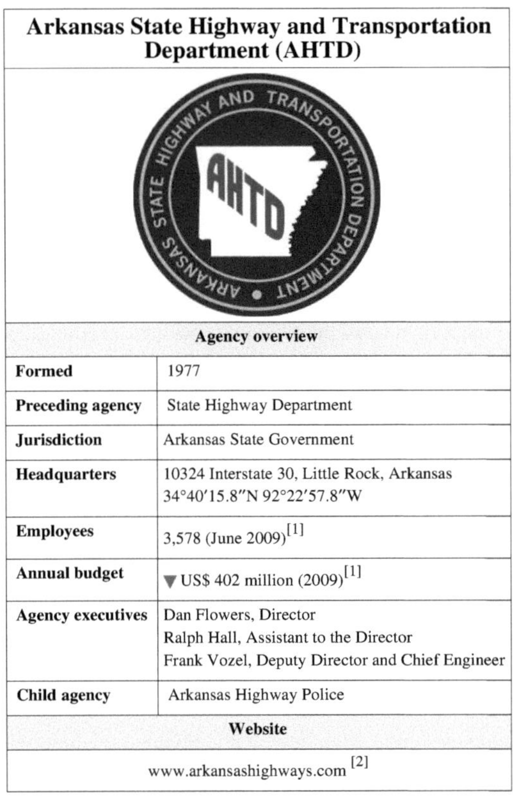</td></tr>
<tr><td colspan="2" align="center">Agency overview</td></tr>
<tr><td>Formed</td><td>1977</td></tr>
<tr><td>Preceding agency</td><td>State Highway Department</td></tr>
<tr><td>Jurisdiction</td><td>Arkansas State Government</td></tr>
<tr><td>Headquarters</td><td>10324 Interstate 30, Little Rock, Arkansas
34°40′15.8″N 92°22′57.8″W</td></tr>
<tr><td>Employees</td><td>3,578 (June 2009)[1]</td></tr>
<tr><td>Annual budget</td><td>▼ US$ 402 million (2009)[1]</td></tr>
<tr><td>Agency executives</td><td>Dan Flowers, Director
Ralph Hall, Assistant to the Director
Frank Vozel, Deputy Director and Chief Engineer</td></tr>
<tr><td>Child agency</td><td>Arkansas Highway Police</td></tr>
<tr><td colspan="2" align="center">Website</td></tr>
<tr><td colspan="2" align="center">www.arkansashighways.com [2]</td></tr>
</table>

The **Arkansas State Highway and Transportation Department** (**AHTD**) is a government department in the U.S. state of Arkansas. Its mission is to provide a safe, efficient, aesthetically pleasing and environmentally sound intermodal transportation system for the user.[3] It manages the state highway system and is actively involved with public transportation systems within the state such as with aid to individual county road systems.[1] Its headquarters are in Little Rock.[4]

History

Originally started as a commission of three elected officials in 1913[5], the agency has since expanded over the years to include a director and five commissioners appointed by the governor as well as its powers to include full oversight into the planning, construction, and maintenance of Arkansas roads.

AHTD and the Nebraska Department of Roads are the only state transportation agencies in the United States whose name still refers to "highways" or "roads"; most others are now known as the "(name of state) Department of Transportation" after the U.S. Department of Transportation. The "Highway" in AHTD's name is largely required by the Arkansas Constitution which created the *Arkansas Highway Commission* as its governing body; the Constitution still calls it the "State Highway Department", but the legislature added "and Transportation" to its name in 1977. Many people in Arkansas continue to call it the "Highway Department" to this day.

Administration

For administrative purposes, AHTD has divided the state of Arkansas into 10 districts[6] supervised by district offices along with 85 county area maintenance headquarters and 31 resident engineer offices located across the state. Most districts cover multiple counties. As a state agency, its central offices are located in Little Rock, which is covered by District 6.[7]

Current projects

- 2007-2010 Statewide Transportation Improvement Program (STIP)
- 2007 Statewide Long-Range Intermodal Transportation Plan
- 2006 Needs Study and Highway Improvement Plan

See also

- U.S. Department of Transportation
- State highways in Arkansas
- Arkansas Transit Association
- *Arkansas Highways*

References

[1] Staff (2009) (PDF). 2009 Facts (http://www.arkansashighways.com/about/2008FACTS.pdf) (Report). Arkansas State Highway and Transportation Department. . Retrieved August 23, 2010.

[2] http://www.arkansashighways.com/

[3] Staff (2007). "Mission Statement" (http://www.arkansashighways.com/about/about_ahtd.aspx). Arkansas State Highway and Transportation Department. . Retrieved August 23, 2010.

[4] Staff (2007). "Contact Us" (http://www.arkansashighways.com/contactus.aspx). Arkansas State Highway and Transportation Department. . Retrieved September 28, 2011.

[5] Scoggin, Robert W. (January 12, 2007). "Arkansas Highway Commission" (http://www.encyclopediaofarkansas.net/encyclopedia/ entry-detail.aspx?entryID=4149). *Encyclopedia of Arkansas History and Culture*. . Retrieved January 27, 2009.

[6] Staff (2007). "Districts" (http://www.arkansashighways.com/district_list.aspx). Arkansas State Highway and Transportation Department. . Retrieved August 23, 2010.

[7] Staff (2007). "District 6" (http://www.arkansashighways.com/district6.aspx). Arkansas State Highway and Transportation Department. . Retrieved August 23, 2010.

External links

- Official website (http://www.arkansashighways.com/)

Arkansas

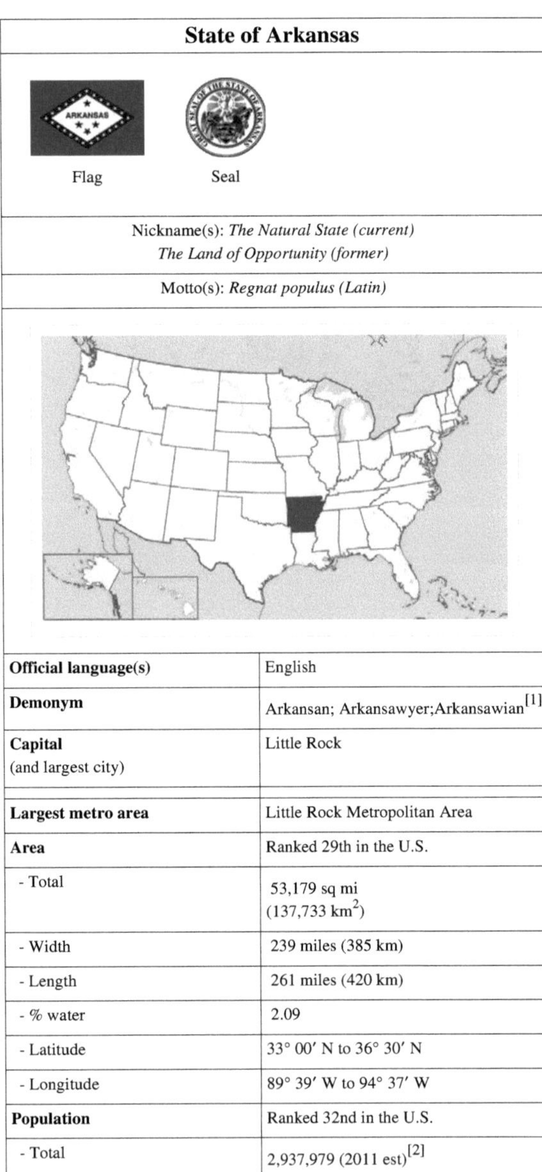

<table>
<tr><td colspan="2" align="center">State of Arkansas</td></tr>
<tr><td align="center">Flag</td><td align="center">Seal</td></tr>
<tr><td colspan="2" align="center">Nickname(s): The Natural State (current)
The Land of Opportunity (former)</td></tr>
<tr><td colspan="2" align="center">Motto(s): Regnat populus (Latin)</td></tr>
</table>

Official language(s)	English
Demonym	Arkansan; Arkansawyer;Arkansawian[1]
Capital (and largest city)	Little Rock
Largest metro area	Little Rock Metropolitan Area
Area	Ranked 29th in the U.S.
- Total	53,179 sq mi (137,733 km^2)
- Width	239 miles (385 km)
- Length	261 miles (420 km)
- % water	2.09
- Latitude	33° 00′ N to 36° 30′ N
- Longitude	89° 39′ W to 94° 37′ W
Population	Ranked 32nd in the U.S.
- Total	2,937,979 (2011 est)[2]

- Density	56.4/sq mi (21.8/km^2) Ranked 34th in the U.S.
Elevation	
- Highest point	Magazine Mountain[3] [4] [5] 2,753 ft (839 m)
- Mean	650 ft (200 m)
- Lowest point	Ouachita River at Louisiana border[4] [5] 55 ft (17 m)
Admission to Union	June 15, 1836 (25th)
Governor	Mike Beebe (D)
Lieutenant Governor	Mark Darr (R)
Legislature	General Assembly
- Upper house	Senate
- Lower house	House of Representatives
U.S. Senators	Mark Pryor (D) John Boozman (R)
U.S. House delegation	3 Republicans, 1 Democrat (list)
Time zone	Central: UTC-6/-5
Abbreviations	AR Ark. US-AR
Website	[www.arkansas.gov www.arkansas.gov]

Arkansas (🔊 [i]/ˈɑrkənsɔː/ *AR-kən-saw*)[6] is a state located in the southern region of the United States. Its name is an Algonquian name of the Quapaw Indians. Arkansas shares borders with six states (N: Missouri; E: Tennessee, Mississippi; S: Louisiana; SW: Texas; W: Oklahoma), and its eastern border is largely defined by the Mississippi River. Its diverse geography ranges from the mountainous regions of the Ozarks and the Ouachita Mountains, which make up the U.S. Interior Highlands, to the eastern lowlands along the Mississippi River. Arkansas is the 29th most extensive and the 32nd most populous of the 50 United States. The capital and most populous city is Little Rock, located in the central portion of the state.

Etymology

The name "Arkansas" derives from the same root as the name for the state of Kansas. The Kansa tribe of Native Americans are closely associated with the Sioux tribes of the Great Plains. The word "Arkansas" itself is a French pronunciation ("Arcansas") of a Quapaw (a related "Kaw" tribe) word, *akakaze*, meaning "land of downriver people" or the Sioux word *akakaze* meaning "people of the south wind". The pronunciation of Arkansas was made official by an act of the state legislature in 1881, after a dispute between two U.S. Senators from Arkansas. One wanted to pronounce the name /ɑrˈkænzəs/ *ar-KAN-zəs* and the other wanted /ˈɑrkənsɔː/ *AR-kən-saw*.[7]

In 2007, the state legislature passed a non-binding resolution declaring the possessive form of the state's name to be *Arkansas's*, which has been followed increasingly by the state government.[8]

Geography

View from the summit of Petit Jean Mountain, nestled in the Arkansas River Valley, from Mather Lodge in Petit Jean State Park.

The Mississippi River forms most of Arkansas's eastern border, except in Clay and Greene counties where the St. Francis River forms the western boundary of the Missouri Bootheel, and in dozens of places where the current channel of the Mississippi has meandered from where it had last been legally specified.[9] Arkansas shares its southern border with Louisiana, its northern border with Missouri, its eastern border with Tennessee and Mississippi, and its western border with Texas and Oklahoma.

Arkansas is a land of lakes and rivers, thick forests and fertile soil. The Arkansas Delta is a flat landscape of rich alluvial soils formed by repeated flooding of the adjacent Mississippi. Farther away from the river, in the southeast portion of the state, the Grand Prairie consists of a more undulating landscape. Both are fertile agricultural areas.

The Delta region is bisected by an unusual geological formation known as Crowley's Ridge. A narrow band of rolling hills, Crowley's Ridge rises from 250 to 500 feet (**unknown operator: u'strong'** m) above the surrounding alluvial plain and underlies many of the major towns of eastern Arkansas.

Northwest Arkansas is part of the Ozark Plateau including the Ozark Mountains, to the south are the Ouachita Mountains, and these regions are divided by the Arkansas River; the southern and eastern parts of Arkansas are called the Lowlands. These mountain ranges are part of the U.S. Interior Highlands region, the only major mountainous region between the Rocky Mountains and the Appalachian Mountains.[10] [11] The highest point in the state is Mount Magazine in the Ouachita Mountains; it rises to 2753 feet (**unknown operator: u'strong'** m) above sea level.

Arkansas is home to many caves, such as Blanchard Springs Caverns. More than 43,000 Native American living, hunting and tool making sites, many of them Pre-Columbian burial mounds and rock shelters, have been catalogued by the State Archeologist. Arkansas is currently the only U.S. state in which diamonds are mined—although by members of the public with primitive digging tools for a small daily fee, not by commercial interests.[12] [13] (near Murfreesboro).

Arkansas is home to many areas protected by the National Park System. These include:[14]

- Arkansas Post National Memorial at Gillett
- Buffalo National River
- Fort Smith National Historic Site
- Hot Springs National Park
- Little Rock Central High School National Historic Site
- Pea Ridge National Military Park
- President William Jefferson Clinton Birthplace Home National Historic Site

Buffalo National River, one of many attractions that give the state's nickname *The Natural State*.

The Trail of Tears National Historic Trail also runs through Arkansas.[14]

Arkansas is home to a dozen Wilderness Areas totaling around 150000 acres (**unknown operator: u'strong'** km^2). These areas are set aside for outdoor recreation and are open to hunting, fishing, hiking, and primitive camping. No mechanized vehicles are allowed in these areas, some of which are rarely visited and can provide a good experience of feeling as if you are the only person to have ever stepped foot there.

Climate

Arkansas generally has a humid subtropical climate, which borders on humid continental in some northern highland areas. While not bordering the Gulf of Mexico, Arkansas is still close enough to this warm, large body of water for it to influence the weather in the state. Generally, Arkansas has hot, humid summers and cold, slightly drier winters. In Little Rock, the daily high temperatures average around 93 °F (**unknown operator: u'strong'** °C) with lows around 73 °F (**unknown operator: u'strong'** °C) in the month of July. In January highs average around 51 °F (**unknown operator: u'strong'** °C) and lows around 32 °F (0 °C). In Siloam Springs in the northwest part of the state, the average high and low temperatures in July are 89 °F (**unknown operator: u'strong'** °C) and 67 °F (**unknown operator: u'strong'** °C) and in January the average high and lows are 44 °F (**unknown operator: u'strong'** °C) and 23 °F (−**unknown operator: u'strong'** °C). Annual precipitation throughout the state averages between about 40 and 60 inches (**unknown operator: u'strong'** and **unknown operator: u'strong'** mm); somewhat wetter in the south and drier in the northern part of the state.[15] Snowfall is common, more so in the north half of the state, which usually gets several snowfalls each winter. This is not only due to its closer proximity to the plains states, but also to the higher elevations found throughout the Ozark and Ouachita mountains. The half of the state south of Little Rock gets less snow, and is more apt to see ice storms, however, sleet and freezing rain are expected throughout the state during the winter months, and can significantly impact travel and day to day life. Arkansas' all time record high is 120 °F (**unknown operator: u'strong'** °C) at Ozark on August 10, 1936; the all time record low is −29 °F (−**unknown operator: u'strong'** °C) at Pond on February 13, 1905.

Arkansas is known for extreme weather. A typical year will see thunderstorms, tornadoes, hail, snow and ice storms. Between both the Great Plains and the Gulf States, Arkansas receives around 60 days of thunderstorms. A few of the most destructive tornadoes in U.S. history have struck the state. While being sufficiently away from the coast to be safe from a direct hit from a hurricane, Arkansas can often get the remnants of a tropical system which dumps tremendous amounts of rain in a short time and often spawns smaller tornadoes.

History

The first European to reach Arkansas was the Spanish explorer Hernando de Soto, a veteran of Pizarro's conquest of Peru who died near Lake Village on the Mississippi River in 1542 after almost a year traversing the southern part of the state in search of gold and a passage to China. Arkansas is one of several U.S. states formed from the territory purchased from Napoleon Bonaparte in the Louisiana Purchase. The early Spanish or French explorers of the state gave it its name, which is probably a phonetic spelling of the Illinois tribe's name for the Quapaw people, who lived downriver from them.[16] Other Native American tribes who lived

Flatside Wilderness Area, Ouachita Mountains, Arkansas

in Arkansas before moving west were the Quapaw, Caddo, and Osage nations. In their forced move westward (under U.S. Indian removal policies), the Five Civilized Tribes inhabited Arkansas during its territorial period.

The Territory of Arkansas[7] was organized on July 4, 1819. On June 15, 1836, the state of Arkansas was admitted to the Union as the 25th state and the 13th slave state. Planters settled in the Delta to cultivate cotton; this was the area of the state where most enslaved African Americans were held. Other areas had more subsistence farmers and mixed farming.

Arkansas played a key role in aiding Texas in its war for independence from Mexico; it sent troops and materials to Texas to help fight the war. The proximity of the city of Washington to the Texas border involved the town in the Texas Revolution of 1835–36. Some evidence suggests Sam Houston and his compatriots planned the revolt in a tavern at Washington in 1834.[17] When the fighting began, a stream of volunteers from Arkansas and the southeastern states flowed through the town toward the Texas battle fields.

When the Mexican-American War began in 1846, Washington became a rendezvous for volunteer troops. Governor Thomas S. Drew issued a proclamation calling on the state to furnish one regiment of cavalry and one battalion of infantry to join the United States Army. Ten companies of men assembled here, where they were formed into the first Regiment of Arkansas Cavalry.

The state developed a cotton culture in the east in lands of the Mississippi Delta. This was where enslaved labor was used most extensively, as planters brought with them or imported slaves from the Upper South. On the eve of the Civil War in 1860, enslaved African Americans numbered 111,115 people, just over 25% of the state's population.[18]

Arkansas refused to join the Confederate States of America until after United States President Abraham Lincoln called for troops to respond to the Confederate attack upon Fort Sumter, South Carolina. The state of Arkansas declared its secession from the Union on May 6, 1861. While not often cited in historical accounts, the state was the scene of numerous small-scale battles during the American Civil War. Arkansans of note who contributed to the Civil War included Confederate Major General Patrick Cleburne. Considered by many to be one of the most brilliant Confederate division commanders of the war, Cleburne was often referred to as "The Stonewall of the West." Also of note was Major General Thomas C. Hindman. A former United States Representative, Hindman commanded Confederate forces at the Battle of Cane Hill and Battle of Prairie Grove.

Under the Military Reconstruction Act, Congress restored Arkansas to the Union in June 1868. The Reconstruction legislature established universal male suffrage while disenfranchising former Confederates (mostly Democrats), a public education system, and other general issues to improve the state and help more of the population. The state came under almost exclusive control of Radical Republicans, (those who emigrated from the North being derided as "carpetbaggers" by ex-Confederates based on allegations of corruption), led by newly elected Governor Powell Clayton, marking a time of great upheaval and racial violence in the state between state militia and the Ku Klux Klan.

In 1874, the Brooks-Baxter War, a political struggle between factions of the Republican Party shook Little Rock and the state governorship. It was settled only when President Ulysses S. Grant ordered Joseph Brooks to disperse his militant supporters.[19]

Following the Brooks-Baxter War, a new state constitution was ratified re-enfranchising former Confederates.

In 1881, the Arkansas state legislature enacted a bill that adopted an official pronunciation of the state's name, to combat a controversy then simmering. (See Law and Government below.)

After Reconstruction, the state began to receive more immigrants and migrants. Chinese, Italian, and Syrian men were recruited for farm labor in the developing Delta region. None of these nationalities stayed long at farm labor; the Chinese especially quickly became small merchants in towns around the Delta. Some early 20th century immigration included people from eastern Europe. Together, these immigrants made the Delta more diverse than the rest of the state. In the same years, some black migrants moved into the area because of opportunities to develop the bottomlands and own their own property. Many Chinese became such successful merchants in small towns that they were able to educate their children at college.[20]

Construction of railroads enabled more farmers to get their products to market. It also brought new development into different parts of the state, including the Ozarks, where some areas were developed as resorts. In a few years at the end of the 19th century, for instance, Eureka Springs in Carroll County grew to 10,000 people, rapidly becoming a tourist destination and the fourth largest city of the state. It featured newly constructed, elegant resort hotels and spas planned around its natural springs, considered to have healthful properties. The town's attractions included horse racing and other entertainment. It appealed to a wide variety of classes, becoming almost as popular as Hot Springs.

In the late 1880s, the worsening agricultural depression catalyzed Populist and third party movements, leading to interracial coalitions. Struggling to stay in power, in the 1890s the Democrats in Arkansas followed other Southern states in passing legislation and constitutional amendments that disfranchised blacks and poor whites. Democrats wanted to prevent their alliance. In 1891 state legislators passed a

Wife and children of a sharecropper in Washington County, Arkansas, c. 1935

requirement for a literacy test, knowing that many blacks and whites would be excluded, at a time when more than 25% of the population could neither read nor write. In 1892 they amended the state constitution to include a poll tax and more complex residency requirements, both of which adversely affected poor people and sharecroppers, and forced them from electoral rolls.

By 1900 the Democratic Party expanded use of the white primary in county and state elections, further denying blacks a part in the political process. Only in the primary was there any competition among candidates, as Democrats held all the power. The state was a Democratic one-party state for decades, until after the Civil Rights Act of 1964 and Voting Rights Act of 1965 were passed.[21]

Between 1905 and 1911, Arkansas began to receive a small migration of German, Slovak, and Irish immigrants. The German and Slovak peoples settled in the eastern part of the state known as the Prairie, and the Irish founded small communities in the southeast part of the state. The Germans were mostly Catholic and the Slovaks were Lutheran. The Irish were mostly Protestant from Ulster.

After the Supreme Court's decision in *Brown* v. *Board of Education of Topeka, Kansas* in 1954, the Little Rock Nine brought Arkansas to national attention when the Federal government intervened to protect African-American students trying to integrate a high school in the Arkansas capital. Governor Orval Faubus ordered the Arkansas National Guard to aid segregationists in preventing nine African-American students from enrolling at Little Rock's Central High School. After attempting three times to contact Faubus, President Dwight D. Eisenhower sent 1000 troops from the active-duty 101st Airborne Division to escort and protect the African-American students as they entered school on September 25, 1957. In defiance of federal court orders to integrate, the governor and city of Little Rock decided to close the high schools for the remainder of the school year. By the fall of 1959, the Little Rock high schools were completely integrated.[22]

Bill Clinton, the 42nd President of the United States, was born in Hope, Arkansas. Before his presidency, Clinton served as the 40th and 42nd Governor of Arkansas, a total of nearly 12 years.

Demographics

Historical populations		
Census	Pop.	%±
1810	1062	—
1820	14273	1244.0%
1830	30388	112.9%
1840	97574	221.1%
1850	209897	115.1%
1860	435450	107.5%
1870	484471	11.3%
1880	802525	65.6%
1890	1128211	40.6%
1900	1311564	16.3%
1910	1574449	20.0%
1920	1752204	11.3%
1930	1854482	5.8%
1940	1949387	5.1%
1950	1909511	−2.0%
1960	1786272	−6.5%
1970	1923295	7.7%
1980	2286435	18.9%
1990	2350725	2.8%
2000	2673400	13.7%
2010	2915918	9.1%
Sources: 1910-2010[23]		

The United States Census Bureau estimates that the population of Arkansas was 2,937,979 on July 1, 2011, a 0.76% increase since the 2010 United States Census.[2]

As of 2006, Arkansas has an estimated population of 2,810,872,[24] which is an increase of 29,154, or 1.1%, from the prior year and an increase of 105,756, or 4.0%, since the year 2000. This includes a natural increase since the last census of 52,214 people (that is 198,800 births minus 146,586 deaths) and an increase due to net migration of 57,611 people into the state. Immigration from outside the United States resulted in a net increase of 21,947 people, and migration within the country produced a net increase of 35,664 people. It is estimated that about 48.8% is male, and 51.2% is female. From 2000 through 2006 Arkansas has had a population growth of 5.1% or 137,472.[25] The population density of the state is 51.3 people per square mile.

According to the 2010 U.S. Census, Arkansas had a population of 2,915,918. In terms of race and ethnicity, the state was 79.5% White (77.5% non-Hispanic White), 14.3% Black or African American, 0.7% American Indian and Alaska Native, 1.2% Asian, 0.2% Native Hawaiian and Other Pacific Islander, 2.8% from Some Other Race, and 1.3% from Two or More Races. Hispanics or Latinos of any race made up 5% of the population.[26]

According to the 2006–2008 American Community Survey,[27] the ten largest ancestry groups in the state are African American (15.5%), Irish (13.6%), German (12.5%), American (11.1%), English (10.3%), French (2.4%), Scotch-Irish (2.1%), Dutch (1.9%), Scottish (1.9%) and Italian (1.7%).

European Americans have a strong presence in the northwestern Ozarks and the central part of the state. African Americans live mainly in the southern and eastern parts of the state. Arkansans of Irish, English and German ancestry are mostly found in the far northwestern Ozarks near the Missouri border. Ancestors of the Irish in the Ozarks were chiefly Scotch-Irish, Protestants from Northern Ireland, the Scottish lowlands and northern England part of the largest group of immigrants from Great Britain and Ireland before the American Revolution. English and Scotch-Irish immigrants settled throughout the backcountry of the South and in the more mountainous areas. Americans of English stock are found throughout the state.[28]

According to the 2006–2008 American Community Survey, 93.8% of Arkansas' population (over the age of five) spoke only English at home. About 4.5% of the state's population spoke Spanish at home. About 0.7% of the state's population spoke any other Indo-European language. About 0.8% of the state's population spoke an Asian language, and 0.2% spoke other languages.

In 2006, Arkansas has a larger percentage of tobacco smokers than the national average, with 24.0% of adults smoking.[29]

Religion

Arkansas, like most other Southern states, is part of the Bible Belt and is predominantly Protestant. The religious affiliations of the people are as follows:[30]

- Christian: 86.0%
  - Protestant: 78.0%
    - Baptist: 39.0%
    - Methodist: 9.0%
    - Pentecostal: 6.0%
    - Church of Christ: 6.0%
    - Assemblies of God: 3.0%
    - Other Protestant: 15.0%
  - Roman Catholic: 7.0%
  - Eastern Orthodox: <1.0%
  - Other Christian: <1.0%
- Non-religious: 14.0%
- Other religions: <1.0%
- Jewish: <1.0%
- Muslim: <1.0%

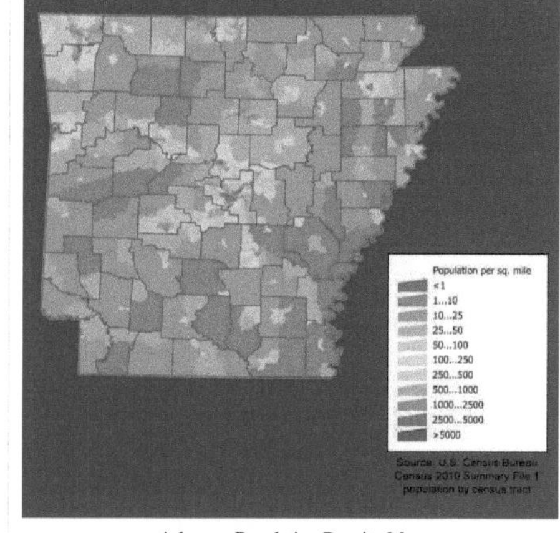

Arkansas Population Density Map

The largest denominations by number of adherents in 2000 were the Southern Baptist Convention with 665,307; the United Methodist Church with 179,383; the Roman Catholic Church with 115,967; and the American Baptist Association with 115,916.[31]

Economy

The quarter for Arkansas, released
October 20, 2003

The state's gross domestic product for 2010 was $103 billion.[32] Its per capita household median income (in current dollars) for 2004 was $35,295, according to the U.S. Census Bureau.[33] The state's agriculture outputs are poultry and eggs, soybeans, sorghum, cattle, cotton, rice, hogs, and milk. Its industrial outputs are food processing, electric equipment, fabricated metal products, machinery, paper products, bromine, and vanadium.

As of August 2011, the state's unemployment rate is 8.3%.[34]

Several global companies are headquartered in the northwest corner of Arkansas, including Wal-Mart (the world's largest public corporation by revenue in 2007),[35] J.B. Hunt and Tyson Foods. This area of the state has experienced an economic boom since the 1970s as a result.

In recent years, automobile parts manufacturers have opened factories in eastern Arkansas to support auto plants in other states.

Tourism is also very important to the Arkansas economy; the official state nickname "The Natural State" was originally created (as "Arkansas Is A Natural") for state tourism advertising in the 1970s, and is still regularly used to this day.

According to Forbes.com[36] Arkansas currently ranks 21st for The Best States for Business, 9th for Business Cost, 40th for Labor, 22nd for Regulatory Environment, 17th for Economic Climate, 9th for Growth Prospects, 34th in Gross Domestic Product, and positive economic change of 3.8% or ranked 22nd.

Largest employers

According to the Arkansas Economic Development Commission, the following are the largest employers in Arkansas by number of in-state employees as of 2010:[37]

Rank	Company
1	State of Arkansas
2	Wal-Mart Stores, Inc.
3	Federal government
4	Tyson Foods, Inc.
5	Baptist Health
6	Acxiom Corporation
7	Arkansas Children's Hospital
8	J.B. Hunt Transport Services, Inc.
9	Sisters of Mercy Health System
10	The Kroger Company
11	Arvest Bank Group, Inc.
12	Community Health Systems, Inc.
13	Simmons Foods, Inc.
14	USA Truck, Inc.
15	Georgia-Pacific Corporation

16	Lowe's Companies, Inc.
17	St. Vincent Infirmary Medical Center
18	FedEx Corporation
19	Entergy Corporation
20	Verizon Communications, Inc. (Cellco Partnership)
21	Union Pacific Railroad Company
22	Dillard's, Inc.
23	PAM Transportation Services, Inc.
24	Pilgrim's Pride Corporation
25	AT&T
26	Baldor Electric Company
27	United Parcel Service
28	St. Bernard's Medical Center
29	Dollar General Corporation
30	Arkansas Blue Cross & Blue Shield
31	Whirlpool Corporation
32	George's, Inc.
33	Husqvarna Home Products
34	Sparks Health System
35	Dassault Aviation Group
36	OK Industries, Inc.
37	Jefferson Regional Medical Center
38	Harps Food Stores, Inc.
39	Walgreen Company
40	University of Central Arkansas
41	Cooper Tire & Rubber Company
42	Washington Regional Medical Center
43	JC Penney Company, Inc.
44	Nucor Corporation
45	The Heritage Company
46	ConAgra Foods, Inc.
47	Riceland Foods, Inc.
48	Home Depot, Inc.
49	McKee Foods Corporation
50	American Greetings Corporation

The University of Arkansas for Medical Sciences, including the UAMS Medical Center, is included in the state of Arkansas' numbers, and would rank in the top ten if listed separately.[37]

Aquaculture

The state also ranks thichannel catfish aquaculture, with about 19200 acres (**unknown operator: u'strong'** km^2) under catfish farming in 2010. The peak of catfish farming in the state was in the year 2002, when 38000 acres (**unknown operator: u'strong'** km^2) were under farming. In 2007, the state's catfish producers generated sales of $71.5 million − 16 percent of the total U.S. market.[38] Arkansas was the first state to develop commercial catfish farms in the late 1950s. The number of catfish farms in the state grew through the 1990s as farmers entered the catfish business as a way to provide additional income during a time of low prices for cotton and soybeans. Banks loaned money to support what they saw as a stable source of revenue.[39]

Taxation

Arkansas imposes a state income tax with six brackets, ranging from 1.0% to 7.0%. The first $9,000 of military pay of enlisted personnel is exempt from Arkansas tax; officers do not have to pay state income tax on the first $6,000 of their military pay. Retirees pay no tax on Social Security, or on the first $6,000 in gain on their pensions along with recovery of cost basis. Residents of Texarkana, Arkansas are exempt from Arkansas income tax; wages and business income earned there by residents of Texarkana, Texas are also exempt. Arkansas's gross receipts (sales) tax and compensating (use) tax rate is currently 6%. The state has also mandated that various services be subject to sales tax collection. They include wrecker and towing services; dry cleaning and laundry; body piercing, tattooing and electrolysis; pest control; security and alarm monitoring; self-storage facilities; boat storage and docking; and pet grooming and kennel services.

A map of Arkansas with county boundaries drawn

Along with the state sales tax, there are more than 300 local taxes in Arkansas. Cities and counties have the authority to enact additional local sales and use taxes if they are passed by the voters in their area. These local taxes have a ceiling or cap; they cannot exceed $25 for each 1% of tax assessed. These additional taxes are collected by the state, which distributes the money back to the local jurisdictions monthly. Low-income taxpayers with a total annual household income of less than $12,000 are permitted a sales tax exemption for electricity usage.

Sales of alcoholic beverages account for added taxes. A 10% supplemental mixed drink tax is imposed on the sale of alcoholic beverages (excluding beer) at restaurants. A 4% tax is due on the sale of all mixed drinks (except beer and wine) sold for "on-premises" consumption. A 3% tax is due on beer sold for off-premises consumption.

Property taxes are assessed on real and personal property; only 20% of the value is used as the tax base.

Transportation

The Interstate Highway system in Arkansas includes

- Interstate 30
- Interstate 40
- Interstate 55
- Interstate 430 Connector Route
- Interstate 440 Connector Route
- Interstate 530 Extension Route
- Interstate 540 Extension Route
- Interstate 630 Connector Route

Also Additional Future Interstate Highways in Arkansas under construction such as:

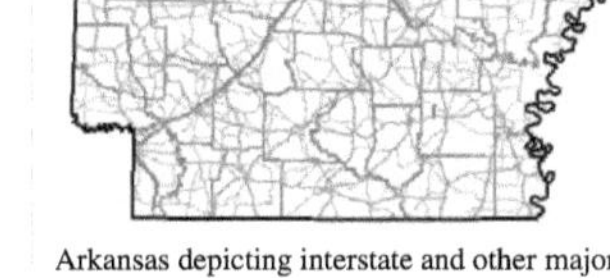

Arkansas depicting interstate and other major highways in 2009.

- 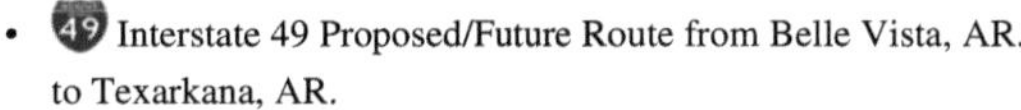 Interstate 49 Proposed/Future Route from Belle Vista, AR. to Texarkana, AR.

- Interstate 69 Future Route From Greenville, MS. to El Dorado, AR.
- Interstate 130 Future Connector Route Texarkana, AR. Area
- Interstate 555 Future Route From Gilmore, AR. to Jonesboro, AR.

There are 20 U.S. Routes in Arkansas, and over 200 Arkansas state highways.

There are four airports with commercial service: Little Rock National Airport, Northwest Arkansas Regional Airport, Fort Smith Regional Airport, and Texarkana Regional Airport, with dozens of smaller airports in the state. The Texas Eagle, an Amtrak passenger train, serves five stations in the state: Walnut Ridge, Little Rock, Malvern, Arkadelphia, and Texarkana. About two dozen railroads carry freight in the state. Public transit and community transport services for the elderly or those with developmental disabilities are provided by agencies such as the Central Arkansas Transit Authority and the Ozark Regional Transit, organizations that are part of the Arkansas Transit Association.

Law and government

The current Governor of Arkansas is Mike Beebe, a Democrat, who was elected on November 7, 2006.[40] [41]

One of Arkansas's U.S. Senators is Democrat Mark Pryor, and the other one is Republican John Boozman. The state has four seats in U.S. House of Representatives. One seat is held by Democrats: Mike Ross (map [42]), and three are held by Republicans: Rick Crawford (politician), (map [43]), Tim Griffin(map [44]), and Steve Womack(map [45]).

Presidential elections results

Year	Republican	Democratic
2008	**58.72%** *638,017*	38.86% *422,310*
2004	**54.31%** *572,898*	44.55% *469,953*
2000	**51.31%** *472,940*	45.86% *422,768*
1996	36.80% *325,416*	**53.74%** *475,171*
1992	35.48% *337,324*	**53.21%** 505,823
1988	**56.37%** *466,578*	42.19% *349,237*
1984	**60.47%** *534,774*	38.29% *338,646*
1980	**48.13%** *403,164*	47.52% *398,041*

1976	34.93% 268,753	**64.94%** *499,614*
1972	**68.82%** *445,751*	30.71% *198,899*
1968*	31.01% *189,062*	30.33% *184,901*
1964	43.41% *243,264*	**56.06%** *314,197*
1960	43.06% *184,508*	**50.19%** *215,049*
*State won by George Wallace of the American Independent Party, at 38.65%, or 235,627 votes		

The Democratic Party holds majority status in the Arkansas General Assembly. A majority of local and statewide offices are also held by Democrats. This is rare in the modern South, where a majority of statewide offices are held by Republicans. Arkansas had the distinction in 1992 of being the only state in the country to give the majority of its vote to a single candidate in the presidential election—native son Bill Clinton—while every other state's electoral votes were won by pluralities of the vote among the three candidates. Arkansas has become more reliably Republican in presidential elections in recent years. The state voted for John McCain in 2008 by a margin of 20 percentage points, making it one of the few states in the country to vote more Republican than it had in 2004. (The others were Louisiana, Tennessee, Oklahoma and West Virginia.)[46] Obama's relatively poor showing in Arkansas was likely due to a lack of enthusiasm from state Democrats following former Arkansas First Lady Hillary Clinton's failure to win the nomination, and his relatively poor performance among rural white voters. However, the Democratic presence remains strong on the state level; in 2006, Democrats were elected to all statewide offices by the voters in a Democratic sweep that included the Democratic Party of Arkansas regaining the governorship, and in 2008, freshman Senator Mark Pryor was re-elected with nearly 80% of the vote against Green candidate Rebekah Kennedy with no Republican opposition.

Most Republican strength lies mainly in the northwestern part of the state, particularly Fort Smith and Bentonville, as well as North Central Arkansas around the Mountain Home area. In the latter area, Republicans have been known to get 90 percent or more of the vote. The rest of the state is more Democratic. Arkansas has only elected two Republicans to the U.S. Senate since Reconstruction, Tim Hutchinson, who was defeated after one term by Mark Pryor and John Boozman, who defeated incumbent Blanche Lincoln. The General Assembly has not been controlled by the Republican Party since Reconstruction and is the fourth most heavily Democratic Legislature in the country, after Massachusetts, Hawaii, and Connecticut. Arkansas was one of just three states among the states of the former Confederacy that sent two Democrats to the U.S. Senate (the others being Florida and Virginia) during the first decade of the 21st century.

Although Democrats have an overwhelming majority of registered voters, Arkansas Democrats tend to be a lot more conservative than their national counterparts, particularly outside Little Rock. Arkansas' Democratic congressman is a member of the Blue Dog Coalition, which tends to be more pro-business, pro-military, and socially conservative than the center-left Democratic mainstream. Reflecting the state's large evangelical population, the state has a strong social conservative bent. Under the Arkansas Constitution Arkansas is a right to work state, its voters passed a ban on same-sex marriage with 75% voting yes, and the state is one of a handful with legislation on its books banning abortion in the event *Roe v. Wade* is ever overturned.

In Arkansas, the lieutenant governor is elected separately from the governor and thus can be from a different political party.

Each officer's term is four years long. Office holders are term-limited to two full terms plus any partial terms before the first full term. Arkansas governors served two-year terms until a referendum lengthened the term to four years, effective with the 1986 general election. Statewide elections are held two years after presidential elections.

Some of Arkansas's counties have two county seats, as opposed to the usual one seat. The arrangement dates back to when travel was extremely difficult in the state. The seats are usually on opposite sides of the county. Though travel is no longer the difficulty it once was, there are few efforts to eliminate the two seat arrangement where it exists, since the county seat is a source of pride (and jobs) to the city involved.

Arkansas is the only state to specify the pronunciation of its name by law (AR-kan-saw).[7]

Article 19 (Miscellaneous Provisions), Item 1 in the Arkansas Constitution is entitled "Atheists disqualified from holding office or testifying as witness," and states that "No person who denies the being of a God shall hold any office in the civil departments of this State, nor be competent to testify as a witness in any Court." However, in 1961, the United States Supreme Court in *Torcaso v. Watkins* (1961), held that a similar requirement in Maryland was unenforceable because it violated the First and Fourteenth Amendments to the US Constitution. The latter amendment, per current precedent, makes the federal Bill of Rights binding on the states. As a result, this provision has not been known to have been enforced in modern times, and it is understood that it would be struck down if challenged in court.

Metropolitan areas

The Little Rock-North Little Rock-Pine Bluff Combined Statistical Area had 877,091 people in the 2010 census. It is the largest in Arkansas.

The Fayetteville-Springdale-Rogers metropolitan area is increasingly important to the state and its economy. The US Census showed the population of the MSA to be 463,204 in 2010 (up from 347,045 in 2000), making it one of the fastest growing areas in the nation.

See also Arkansas Metropolitan Areas.

Cities with 10,000 or More Residents as of 2010

Rank	City	2010 Pop.	Region
1.	Little Rock, Arkansas	193,524	Central
2.	Fort Smith, Arkansas	86,209	Northwest
3.	Fayetteville, Arkansas	73,580	Northwest
4.	Springdale, Arkansas	69,797	Northwest
5.	Jonesboro, Arkansas	67,263	Northeast
6.	North Little Rock, Arkansas	62,304	Central
7.	Conway, Arkansas	58,908	Central
8.	Rogers, Arkansas	55,964	Northwest
9.	Pine Bluff, Arkansas	49,083	Southeast
10.	Bentonville, Arkansas	35,301	Northwest
11.	Hot Springs, Arkansas	35,193	Southwest
12.	Texarkana, Arkansas	30,681	South
13.	Benton, Arkansas	29,919	Central
14.	Sherwood, Arkansas	29,523	Central
15.	Jacksonville, Arkansas	28,364	Central
16.	Russellville, Arkansas	27,920	Northwest
17.	Bella Vista, Arkansas	26,461	Northwest
18.	West Memphis, Arkansas	26,245	Northeast
19.	Paragould, Arkansas	26,113	Northeast
20.	Cabot, Arkansas	23,776	Central

21.	Searcy, Arkansas	22,858	Central
22.	Van Buren, Arkansas	22,791	Northwest
23.	El Dorado, Arkansas	18,884	South-Central
24.	Maumelle, Arkansas	17,163	Central
25.	Bryant, Arkansas	16,688	Central
26.	Blytheville, Arkansas	15,620	Northeast
27.	Forrest City, Arkansas	15,371	Northeast
28.	Siloam Springs, Arkansas	15,039	Northwest
29.	Harrison, Arkansas	12,943	Northwest
30.	Mountain Home, Arkansas	12,448	Northwest
31.	Marion, Arkansas	12,345	Northeast
32.	Helena-West Helena, Arkansas	12,282	Northeast
33.	Camden, Arkansas	12,183	Southwest
34.	Magnolia, Arkansas	11,577	Southwest
35.	Arkadelphia, Arkansas	10,714	Southwest
36.	Malvern, Arkansas	10,318	Southwest
37.	Batesville, Arkansas	10,248	Northeast
38.	Hope, Arkansas	10,095	Southwest

These population numbers are according to the 2010 US Census.

Cities and towns

Names in **bold** have populations greater than 20,000.

Little Rock is Arkansas' capital and most populous city

Fort Smith

Fayetteville

Camden

Magnolia

- Alma
- Alpena
- Arkadelphia
- Augusta
- Avoca
- Batesville
- Bay
- Beech Grove
- Beebe
- Bella Vista
- Bellefonte
- **Benton**
- **Bentonville**
- Bergman
- Harrison
- Horatio
- Helena-West Helena
- Hope
- **Hot Springs**
- Hoxie
- Hughes, Arkansas
- Jasper
- **Jacksonville**
- **Jonesboro**
- Kingsland
- Lafe
- Lake City
- Lake Village
- Siloam Springs
- Smackover
- **Springdale**
- South Lead Hill
- Stuttgart
- Summit
- Swifton
- **Texarkana**
- Trumann
- Tuckerman
- Valley Springs
- **Van Buren**
- Waldron
- Walnut Ridge

- Berryville
- Biggers
- Black Oak
- Black Rock
- Blytheville
- Bono
- Booneville
- Bryant
- Brookland
- **Cabot**
- Camden
- Caraway
- Cash
- **Conway**
- Clinton
- College City
- Corning
- Cotter
- Crossett
- Dardanelle
- Delaplaine
- Dell
- DeQueen
- De Witt
- Diamond City
- Earle
- **El Dorado**
- Egypt
- Eureka Springs
- Everton
- **Fayetteville**
- Forrest City
- **Fort Smith**
- Flippin
- Gassville
- Greenwood
- Grubbs
- Gurdon

- Leachville
- Lead Hill
- **Little Rock**
- Lonoke
- Lockesburg
- Magnolia
- Malvern
- Marion
- Marked Tree
- Maumelle
- Mena
- Midway
- Monticello
- Morrilton
- Monette
- Mountain Home
- Mountain View
- Natural Steps
- Newport
- **North Little Rock**
- Omaha
- Osceola
- Otwell
- **Paragould**
- Parkin
- **Pine Bluff**
- Piggott
- Pleasant Hill
- Pocahontas
- Pyatt
- Ravenden
- Ravenden Springs
- Rea Valley
- Rison
- **Rogers**
- **Russellville**
- **Searcy**
- **Sherwood**
- Sheridan

- Warren
- **West Memphis**
- Weiner
- Western Grove
- Wynne
- Yardelle
- Yellville
- Zinc

Education

Public school districts

- List of school districts in Arkansas

Centers of research

- National Center for Toxicological Research
- University of Arkansas Division of Agriculture

Colleges and universities

- Arkansas Baptist College
- Arkansas State University System
  - Arkansas State University – Jonesboro
  - Arkansas State University Beebe
  - Arkansas State University - Mountain Home
- Arkansas Tech University
- Central Baptist College
- Crowley's Ridge College
- Harding University
- Henderson State University
- Hendrix College
- John Brown University
- Lyon College
- Northwest Arkansas Community College
- Ouachita Baptist University
- Ozarka College
- Philander Smith College
- Southern Arkansas University
- University of Arkansas System
  - University of Arkansas at Fayetteville
  - University of Arkansas at Fort Smith
  - University of Arkansas at Little Rock
  - University of Arkansas for Medical Sciences
  - University of Arkansas at Monticello
  - University of Arkansas at Pine Bluff
  - University of Arkansas Community College at Batesville
  - University of Arkansas Community College at Morrilton
- University of Central Arkansas
- University of the Ozarks
- Williams Baptist College

University of Arkansas, Fayetteville.

UAMS is the flagship health education institution of the state.

Notable residents

- Joey Lauren Adams
- Homer Martin Adkins
- Bill Alexander
- Dale Alford
- Kris Allen
- Maya Angelou
- Beryl Anthony
- Morris Arnold
- Richard S. Arnold
- Harry Ashmore
- Wayne H. Babbitt
- Carl Edward Bailey
- Gilbert Baker
- Richard Barclay
- Daisy Bates
- Garland E. Bayliss
- Mike Beebe
- Bruce Bennett
- Robert Marion Berry
- Edwin R. Bethune
- Len Blaylock
- Brian Bohrer
- Fay Boozman
- John Boozman
- Vickey Boozman
- Henry M. Britt
- Maurice L. Britt
- Lou Brock
- Charles Hillman Brough
- Frank Broyles
- Dee Brown
- Helen Gurley Brown
- Paul "Bear" Bryant
- Dale Bumpers
- Preston Bynum
- Glen Campbell
- Hattie Caraway
- Johnny Cash
- William Lee Cazort
- Francis Cherry
- Norris Church (Barbara Jean Davis)
- Wesley Clark
- Jerry Climer
- Bill Clinton
- Chelsea Clinton
- Hillary Rodham Clinton

- L. L. Clover
- Ken Coon
- James Cotten
- Maud Crawford
- Fulham Davies
- Ronald N. Davies
- Lynn A. Davis
- "Dizzy" Dean
- Elizabeth R. Deans
- Bill Dickey
- Jay Dickey
- Beth Ditto
- David O. Dodd
- Jimmy Driftwood
- Duggar family
- Jimmy Dykes
- G. Thomas Eisele
- Orval Faubus
- Derek Fisher
- John Gould Fletcher
- Connie Franklin
- Woody Freeman
- J. William Fulbright
- Junius Marion Futrell
- Augustus Garland
- Tyson Gay
- David Delano Glover
- Robert W. Glover
- Nathan Green Gordon
- Leon Griffith
- John Grisham
- E. Lynn Harris
- Oren Harris
- John Paul Hammerschmidt
- Ronnie Hawkins
- Brooks Hays
- Jim Holt
- Ernest Hemingway
- Kim Hendren
- Peyton Hillis
- Harlan Holleman
- Mike Huckabee
- Johnnie Bryan Hunt
- Torii Hunter
- Walter E. Hussman, Jr.
- Walter E. Hussman, Sr.
- Asa Hutchinson

- Jeremy Hutchinson
- Tim Hutchinson
- Joe Jackson
- Keith Jackson
- Travis Jackson
- James Douglas "Justice Jim" Johnson
- Joe Johnson
- John H. Johnson
- Virginia Morris Johnson
- Jerry Jones
- Scott Joplin
- Jim Keet
- George Kell
- Larry Lacewell
- Alan Ladd
- Benjamin T. Laney
- Marjorie Lawrence
- Amy Lee
- Cliff Lee
- Blanche Lambert Lincoln
- Jim Lindsey
- Sonny Liston
- A. Lynn Lowe
- Josh Lucas
- Douglas MacArthur
- Charlotte Mailliard
- Mark Martin
- John Little McClellan
- Hayes McClerkin
- James S. McDonnell
- James McDougall
- Darren McFadden
- Sid McMath
- Thomas Chipman McRae
- John Ellis Martineau
- C.W. Melson
- John E. Miller
- Ada Mills
- Wilbur Mills
- Martha Beall Mitchell
- Ben Moody
- Justin Moore
- Jonathan Moreira
- Sheffield Nelson
- Joe Nichols
- Catherine Dorris Norrell
- William F. Norrell

- Clyde E. Palmer
- Robert Palmer
- Harvey Parnell
- Bobby Petrino
- Judy Petty Wolf
- Albert Pike
- Scottie Pippen
- Carolyn Pollan
- Odell Pollard
- Charles Portis
- Dick Powell
- A. T. Powers
- David H. Pryor
- Mark Pryor
- Joe Purcell
- James B. Reed
- Pratt Remmel
- Willis Ricketts
- Thomas Robb
- Brooks Robinson
- Joseph T. Robinson
- Tommy F. Robinson
- Winthrop Rockefeller
- Winthrop Paul Rockefeller
- Mike Ross
- Monroe Schwarzlose
- Shaffer Smith
- Jim L. Smithson
- John W. Snyder
- Jefferson W. Speck
- Sir Henry Morton Stanley
- Mary Steenburgen
- Edward Durell Stone
- Pat Summerall
- Barry Switzer
- Boyd Anderson Tackett
- Reece "Goose" Tatum
- Jermain Taylor
- Tom Jefferson Terral
- David D. Terry
- Jerry Thomasson
- Billy Bob Thornton
- James William Trimble
- Thomas Clark Trimble, III
- Leona Troxell
- Jim Guy Tucker
- Don Tyson

- Steve Womack
- C. Vann Woodward
- Arky Vaughan
- Ehron VonAllen
- Sam Walton
- Joseph H. Weston
- Frank D. White
- Barry Williamson
- Effiegene Locke Wingo
- Otis Wingo
- James Lee Witt
- Travis Wood
- Archibald Yell
- Eden Wood

See also

- List of National Register of Historic Places in Arkansas
- List of people from Arkansas
- List of places in Arkansas
- U.S. state

References

[1] "Arkansawyer definition" (http://encarta.msn.com/dictionary_1861695659_1861695659/prevpage.html). *Arkansawyer definition*. 18 May
2010. .

[2] "Annual Estimates of the Resident Population for the United States, Regions, States, and Puerto Rico: April 1, 2010 to July 1, 2011" (http://
www.census.gov/popest/data/state/totals/2011/tables/NST-EST2011-01.csv) (CSV). *2011 Population Estimates*. United States Census
Bureau, Population Division. December 2011. . Retrieved December 21, 2011.

[3] "Mag" (http://www.ngs.noaa.gov/cgi-bin/ds_mark.prl?PidBox=FG1888). *NGS data sheet*. U.S. National Geodetic Survey. . Retrieved
October 20, 2011.

[4] "Elevations and Distances in the United States" (http://egsc.usgs.gov/isb/pubs/booklets/elvadist/elvadist.html). United States
Geological Survey. 2001. . Retrieved October 21, 2011.

[5] Elevation adjusted to North American Vertical Datum of 1988.

[6] Jones, Daniel. (1997) *English Pronouncing Dictionary*, 15th ed. Cambridge University Press. ISBN 0-521-45272-4

[7] The name Arkansas has been pronounced and spelled in a variety of fashions. The region was organized as the Territory of Arkansaw on July
4, 1819, but the territory was admitted to the United States as the state of Arkansas on June 15, 1836. The name was historically /ˈɑrkənsɔː/,
/ɑrˈkænzəs/, and several other variants. In 1881, the Arkansas General Assembly passed the following concurrent resolution, now Arkansas
Code 1-4-105 (official text (http://www.arkleg.state.ar.us/assembly/ArkansasCode/0/1-4-105.htm)):

> Whereas, confusion of practice has arisen in the pronunciation of the name of our state and it is deemed
> important that the true pronunciation should be determined for use in oral official proceedings.
>
> And, whereas, the matter has been thoroughly investigated by the State Historical Society and the
> Eclectic Society of Little Rock, which have agreed upon the correct pronunciation as derived from
> history, and the early usage of the American immigrants.
>
> Be it therefore resolved by both houses of the General Assembly, that the only true pronunciation of the
> name of the state, in the opinion of this body, is that received by the French from the native Indians and
> committed to writing in the French word representing the sound. It should be pronounced in three (3)
> syllables, with the final "s" silent, the "a" in each syllable with the Italian sound, and the accent on the
> first and last syllables. The pronunciation with the accent on the second syllable with the sound of "a" in
> "man" and the sounding of the terminal "s" is an innovation to be discouraged.

Citizens of the state of Kansas often pronounce the Arkansas River as /ɑrˈkænzəsˈrɪvər/, in a manner similar to the common pronunciation of the name of their state.

[8] Gambrell, John. "Senate gives support to possessive form of Arkansas." (http://www2.arkansasonline.com/news/2007/mar/13/senate-gives-support-possessive-form-arkansas/) *Arkansas Democrat Gazette*, March 13, 2007.

[9] Arkansas State Boundaries (http://www.encyclopediaofarkansas.net/encyclopedia/entry-detail.aspx?entryID=2546) from the Encyclopedia of Arkansas (http://www.encyclopediaofarkansas.net/)

[10] "Managing Upland Forests of the Midsouth" (http://www.srs.fs.usda.gov/4106/about/HotSpringsOffice.htm). United States Forestry Service. . Retrieved 2007-10-13.

[11] "A Tapestry of Time and Terrain: The Union of Two Maps – Geology and Topography" (http://tapestry.usgs.gov/physiogr/physio.html). United States Geological Survey. . Retrieved 2007-10-13.

[12] "Crater of Diamonds: History of diamonds, diamond mining in Arkansas" (http://www.craterofdiamondsstatepark.com/history/). Craterofdiamondsstatepark.com. . Retrieved 2010-07-30.

[13] "US Diamond Mines – Diamond Mining in the United States" (http://geology.com/gemstones/united-states-diamond-production.shtml). Geology.com. . Retrieved 2010-07-30.

[14] "Arkansas" (http://www.nps.gov/state/ar). National Park Service. . Retrieved 2008-07-15.

[15] Average Annual Precipitation – Arkansas (http://www.ocs.orst.edu/pub/maps/Precipitation/Total/States/AR/ar.gif). Spatial Climate Analysis Service, Oregon State University. Published 2000. Last Retrieved 2007-10-26.

[16] "Linguist list 14.4" (http://listserv.linguistlist.org/cgi-bin/wa?A2=ind0302b&L=ads-l&P=7800). Listserv.linguistlist.org. 2003-02-11. . Retrieved 2010-07-30.

[17] Taylor, Jim. "Old Washington State Park Conserves Town's Heyday" (http://www.arkansasmediaroom.com/news-releases/listings/display.asp?id=165). .

[18] Historical Census Browser, 1860 US Census, University of Virginia (http://fisher.lib.virginia.edu/collections/stats/histcensus/php/state.php). Retrieved March 21, 2008.

[19] "Brooks-Baxter War – Encyclopedia of Arkansas" (http://www.encyclopediaofarkansas.net/encyclopedia/entry-detail.aspx?entryID=2276). . Retrieved 2007-08-24.

[20] William D. Baker, *Minority Settlement in the Mississippi River Counties of the Arkansas Delta, 1870–1930*, Arkansas Preservation Commission (http://www.arkansaspreservation.org/pdf/publications/Minority_Settlement.pdf), accessed 14 May 2008

[21] "White Primary" System Bars Blacks from Politics – 1900", *The Arkansas News*, Old State House, Spring 1987, p.3. Retrieved March 22, 2008. (http://www.oldstatehouse.com/educational_programs/classroom/arkansas_news/detail.asp?id=800&issue_id=36&page=3)

[22] "Little Rock Nine – Encyclopedia of Arkansas" (http://www.encyclopediaofarkansas.net/encyclopedia/entry-detail.aspx?search=1&entryID=723). . Retrieved 2007-08-24.

[23] "Resident Population Data - 2010 Census" (http://2010.census.gov/2010census/data/apportionment-pop-text.php). 2010.census.gov. . Retrieved 2012-01-26.

[24] "Annual Estimates of the Population for the United States and States, and for Puerto Rico: April 1, 2000 to July 1, 2005" (http://www.census.gov/popest/states/tables/NST-EST2005-01.csv) (CSV). *2005 Population Estimates*. U.S. Census Bureau, Population Division. June 21, 2006. . Retrieved November 15, 2006.

[25] "Arkansas QuickFacts from the US Census Bureau" (http://quickfacts.census.gov/qfd/states/05000.html). .

[26] "American FactFinder" (http://factfinder2.census.gov/faces/tableservices/jsf/pages/productview.xhtml?pid=DEC_10_PL_QTPL&prodType=table). Factfinder2.census.gov. 2010-10-05. . Retrieved 2012-01-26.

[27] American FactFinder, United States Census Bureau. "Arkansas – Selected Social Characteristics in the United States: 2006–2008" (http://factfinder.census.gov/servlet/ADPTable?_bm=y&-geo_id=04000US05&-qr_name=ACS_2008_3YR_G00_DP3YR2&-ds_name=ACS_2008_3YR_G00_&-_lang=en&-_sse=on). Factfinder.census.gov. . Retrieved 2010-07-30.

[28] David Hackett Fischer, *Albion's Seed: Four British Folkways in America*, New York: Oxford University Press, 1989, pp.633–639

[29] CDC's State System – State Comparison Report Cigarette Use (Adults) – BRFSS (http://apps.nccd.cdc.gov/StateSystem/stateSystem.aspx?selectedTopic=100&selectedMeasure=1000&dir=epi_report&ucName=UCSummary&year=2006&excel=htmlTable&submitBk=y) for 2006, lists the state as having 23.7% smokers. The national average is 20.8% according to Cigarette Smoking Among Adults --- United States, 2006 (http://www.cdc.gov/mmwr/preview/mmwrhtml/mm5644a2.htm) article in the CDC's Morbidity and Mortality Weekly Report.

[30] "American Religious Identification Survey, 2001" (http://www.gc.cuny.edu/faculty/research_briefs/aris/key_findings.htm). Gc.cuny.edu. . Retrieved 2010-07-30.

[31] "The Association of Religion Data Archives I Maps & Reports" (http://www.thearda.com/mapsReports/reports/state/05_2000.asp). Thearda.com. . Retrieved 2010-07-30.

[32] "GDP by State" (http://greyhill.com/gdp-by-state). Greyhill Advisors. . Retrieved 23 September 2011.

[33] Arkansas QuickFacts (http://quickfacts.census.gov/qfd/states/05000.html) from the US Census Bureau

[34] "Local Area Unemployment Statistics" (http://www.bls.gov/web/laus/laumstch.htm). BLS. . Retrieved 23 September 2011.

[35] Staff Writer. " Fortune Global 500 (http://money.cnn.com/magazines/fortune/global500/2007/)." *CNN/Fortune*. 2007. Retrieved on November 8, 2007.

[36] "Table: The Best States For Business" (http://www.forbes.com/2007/07/10/washington-virginia-utah-biz-cz_kb_0711bizstates-table.html). Forbes.com. July 11, 2007. . Retrieved 2010-07-30.

[37] " Arkansas' Largest Employers -- 2010 (http://www.arkansasedc.com/media/118399/arkansas' largest employers.pdf)" (August 2010).
     Arkansas Economic Development Commission, Research Division.
[38] (http://usda.mannlib.cornell.edu/usda/nass/CatfProd//2000s/2002/CatfProd-02-07-2002.pdf) from USDA NASS
[39] (http://www.arfb.com/news_information/ark_agri/2008v5i2/catfish.aspx) Arkansas Farm Bureau Federation
[40] "Winners in '06 Governors races" (http://archive.stateline.org/weekly/Stateline.org-Weekly-Original-Content-2006-11-06.pdf) (PDF). .
     Retrieved 2010-07-30.
[41] "Arkansas.gov Administration page for Governor" (http://dwe.arkansas.gov/GenIfo/administration.html). Dwe.arkansas.gov.
     2007-03-16. . Retrieved 2010-07-30.
[42] http://nationalatlas.gov/printable/images/preview/congdist/ar04_109.gif
[43] http://nationalatlas.gov/printable/images/preview/congdist/ar01_109.gif
[44] http://nationalatlas.gov/printable/images/preview/congdist/ar02_109.gif
[45] http://nationalatlas.gov/printable/images/preview/congdist/ar03_109.gif
[46] http://upload.wikimedia.org/wikipedia/commons/5/54/Election-state-04-08.png

Further reading

- Blair, Diane D. & Jay Barth *Arkansas Politics & Government: Do the People Rule?* (2005)
- Deblack, Thomas A. *With Fire and Sword: Arkansas, 1861–1874* (2003)
- Donovan, Timothy P. and Willard B. Gatewood Jr., eds. *The Governors of Arkansas* (1981)
- Dougan, Michael B. *Confederate Arkansas* (1982),
- Duvall, Leland. ed., *Arkansas: Colony and State* (1973)
- Fletcher, John Gould. *Arkansas* (1947)
- Hamilton, Peter Joseph. *The Reconstruction Period* (http://books.google.com/books?ie=UTF-8&
  vid=LCCN04001664&id=dvMZ_YNpYrYC&pg=PP16&lpg=PP16&dq=peter+joseph+hamilton+
  reconstruction+period) (1906), full length history of era; Dunning School approach; 570 pp; ch 13 on Arkansas
- Hanson, Gerald T. and Carl H. Moneyhon. *Historical Atlas of Arkansas* (1992)
- Key, V. O. *Southern Politics* (1949)
- Kirk, John A., *Redefining the Color Line: Black Activism in Little Rock, Arkansas, 1940–1970* (2002).
- McMath, Sidney S. *Promises Kept* (2003)
- Moore, Waddy W. ed., *Arkansas in the Gilded Age, 1874–1900* (1976).
- Peirce, Neal R. *The Deep South States of America: People, Politics, and Power in the Seven Deep South States*
  (1974)\
- Thompson, Brock. *The Un-Natural State: Arkansas and the Queer South* (2010)
- Thompson, George H. *Arkansas and Reconstruction* (1976)
- Whayne, Jeannie M. et al. *Arkansas: A Narrative History* (2002)
- Whayne, Jeannie M. *Arkansas Biography: A Collection of Notable Lives* (2000)
- White, Lonnie J. *Politics on the Southwestern Frontier: Arkansas Territory, 1819–1836* (1964)
- Williams, C. Fred. ed. *A Documentary History Of Arkansas* (2005)
- WPA., *Arkansas: A Guide to the State* (1941)

External links

- Arkansas.gov (http://portal.arkansas.gov/Pages/default.aspx) (Official Website for the State of Arkansas)
- Arkansas State Code (the state statutes of Arkansas) (http://www.arkleg.state.ar.us/data/ar_code.asp)
- Arkansas State Databases (http://wikis.ala.org/godort/index.php/Arkansas) – Annotated list of searchable
  databases produced by Arkansas state agencies and compiled by the Government Documents Roundtable of the
  American Library Association.
- Arkansas State Facts (http://www.ers.usda.gov/statefacts/ar.htm)
- Official State tourism website (http://www.arkansas.com)
- The Encyclopedia of Arkansas History & Culture (http://www.encyclopediaofarkansas.net/)

- Energy & Environmental Data for Arkansas (http://tonto.eia.doe.gov/state/state_energy_profiles. cfm?sid=AR)
- U.S. Census Bureau (http://quickfacts.census.gov/qfd/states/05000.html)
- 2000 Census of Population and Housing for Arkansas (http://www2.census.gov/census_2000/datasets/ demographic_profile/Arkansas/2kh05.pdf), U.S. Census Bureau
- USGS real-time, geographic, and other scientific resources of Arkansas (http://www.usgs.gov/state/state. asp?State=AR)
- Arkansas Summer Camps (http://www.summercampsus.com/arkansas-summer-camps/)
- Arkansas Shakespeare Theatre (http://www.arkshakes.com)
- Arkansas (http://ballotpedia.org/wiki/index.php/Arkansas) at Ballotpedia
- Arkansas (http://judgepedia.org/index.php/Arkansas) at Judgepedia
- Arkansas (http://sunshinereview.org/index.php/Arkansas) at Sunshine Review
- Arkansas (http://www.dmoz.org/Regional/North_America/United_States/Arkansas/) at the Open Directory Project

Article Sources and Contributors

Arkansas Highway 204 *Source*: http://en.wikipedia.org/w/index.php?title=Arkansas_Highway_204 *Contributors*: Appraiser, Brandonrush, Ltljltlj, NE2, Poccil, Rschen7754, SPUI, US 71, 1 anonymous edits

Bentonville Municipal Airport *Source*: http://en.wikipedia.org/w/index.php?title=Bentonville_Municipal_Airport *Contributors*: Hmains, Zyxw

Benton County, Arkansas *Source*: http://en.wikipedia.org/w/index.php?title=Benton_County%2C_Arkansas *Contributors*: AXRL, Acntx, Ark30inf, AshyLarry, Backspace, BartekChom, Basketballer1042, Bogger, Catbar, Danielba894, DoxTxob, Dravecky, Efghij, Elephant200, Gobonobo, Googa-gook, GrahamHardy, Hall Monitor, HennessyC, Hephaestos, Hu12, Hyperfast, Jafir, Jaknouse, Jessicashabatura, Jfairbanks, Jmlk17, Johnpacklambert, Kenner116, Ksademap, LouI, MrBill47, NE2, NWAJason, NawlinWiki, Neko-chan, NetherlandishYankee, Niceguyedc, Nyttend, Ohnoitsjamie, Omnedon, Orlady, Pearle, Phyllis1753, Ram-Man, Rich Farmbrough, SD5, Saxbryn, Smallbones, Susan12, Swtang, Template namespace initialisation script, The Moose, TheCatalyst31, Theduckman1763, Thunderstruck401, Tim!, Twp, Voyager, WillC, Worldenc, Yx7791, Zoe, Zoicon5, 33 anonymous edits

Siloam Springs, Arkansas *Source*: http://en.wikipedia.org/w/index.php?title=Siloam_Springs%2C_Arkansas *Contributors*: Acntx, Ahoerstemeier, Anlace, Apparition11, Bishop.nate, Brian1979, Catbar, ChrisCork, Chriscapehart, DVdm, Daven200520, Dekimasu, Dwo, Encycloar, Epolk, Esrever, Hephaestos, Hmains, Hushpuckena, JCarriker, John Cardinal, Johnpacklambert, Joyous!, Jrellis77, Kaltenmeyer, Kroma, Lpkm79, Meggar, Meredith Bergstrom, Mild Bill Hiccup, Nyttend, Ofus, Pfly, Pinethicket, Ram-Man, Redf0x, Rich Farmbrough, Rjwilmsi, SFB2388, Samsara, Scwlong, SpikeToronto, Templerc, Tim Morgan, Unitedbride, Woohookitty, 67 anonymous edits

Northwest Arkansas Regional Airport *Source*: http://en.wikipedia.org/w/index.php?title=Northwest_Arkansas_Regional_Airport *Contributors*: Allstar86, Andrew Kurish, Audude08, Bobo192, Buffalutheran, Burgundavia, Cardsplayer4life, Cashier freak, Charmedaddict, Chikinsawsage, Choster, D.c.camero, Daltnpapi4u, Dbchip, Deli nk, Drdisque, FCYTravis, Gittinsj, Hawaiian717, Hmains, Ibagli, Ikara, Insanedevilray, Jerrykme, Jetset59, K50 Dude, KnightRider, LAAzerwolf, Lightmouse, LurkingInChicago, Lvkewlkid, Marketing1722, Mnsourcer, Pantherclaw, PikDig, Quackslikeaduck, Quidam65, RBBrittain, RedWolf, Redlegsfan21, RightSideNov, Rjwilmsi, Rmcclen, SchuminWeb, Senjuto, SmthManly, Snoozlepet, Sox23, Spendeau, Tabletop, Thadius856, Tinlinkin, Vegaswikian, WhisperToMe, Zscout370, Zyxw, 101 anonymous edits

Bannered routes of U.S. Route 71 *Source*: http://en.wikipedia.org/w/index.php?title=Bannered_routes_of_U.S._Route_71 *Contributors*: BD2412, Brandonrush, Cappicard, Chowbok, DandyDan2007, Dough4872, Fredddie, Imzadi1979, Kinu, LilHelpa, Ljthefro, Mapsax, Mcdonaat, Mickeydeesium-118, Morriswa, NE2, Rich Farmbrough, Rschen7754, Scott5114, TwinsMetsFan, US 71, Ulric1313, Will Beback Auto, 9 anonymous edits

Arkansas State Highway and Transportation Department *Source*: http://en.wikipedia.org/w/index.php?title=Arkansas_State_Highway_and_Transportation_Department *Contributors*: Alansohn, Brandonrush, Docu, Eastlaw, Fitzaubrey, Fry1989, GrahamHardy, Imzadi1979, Johnpacklambert, Overpush, RBBrittain, Secondarywaltz, TenPoundHammer, WhisperToMe, Yadyn, 5 anonymous edits

Arkansas *Source*: http://en.wikipedia.org/w/index.php?title=Arkansas *Contributors*: 1.21 jigwatts, 11tas, 16@r, 208.144.114.xxx, 72Dino, 75th Trombone, 90, A8UDI, ALifeMoreHerbal, ARStudent2009, Abcabc123, Acather96, Achangeisasgoodasa, Acntx, Acroterion, Adam 1212, Addison0426, Addshore, AdjustShift, Adrian.benko, Aesopos, AeturnalNarcosis, Aeusoes1, Agcshm, Aghalee2008, Ahassan05, Ahoerstemeier, Airtuna08, Aitias, Aivazovsky, Akuyume, Al Silonov, Alanjackson10, Alansohn, Alex Arnold, Alex43223, AlexiusHoratius, Alexwcovington, Allen4names, Allstarecho, Alrees, Altenmann, Amaamaddq, Amazonien, Ambersharae, Amikeco, Amiodarone, Amire80, Amnesiac101, Analogdrift, Andy Marchbanks, Andy120290, Andycjp, Angr, Animum, Antandrus, Ante Aikio, Anynabiddle, Apalmer03, Arattorney, Archmagi1, Ark30inf, ArkansasRazorback, ArkansasTraveler, Arkansasgirl2009, Arkresi, ArmadilloFromHell, Arx Fortis, Arxiloxos, Asdfghjkl1222222, Ashes fall, AshleyMotown, Astynax, AuburnPilot, Aude, Auntof6, Axl, Ayamtelur, Aztom2, B1157, BD2412, BSveen, Baa, Badagnani, Badgernet, Balcer, Barek, Barnabypage, Barrabhoy, BartLIV, Basquetteur, Bbalgirl10124, Beedubaya, Beetstra, Beezer137, Behun, Beland, Belovedfreak, BenBaker, BenH, Bender235, Benhealy, Benjh40, Benstown, Betacommand, Bgnib, Bidgee, BigrTex, Billim1, Billrnyc, Billy Hathorn, BirdValiant, Blanchardb, Blue520, Bmv 1978, Bob2009007, Bobianite, Bobjohn69, Bobo192, Bomac, Bongwarrior, Bookjunky, Bookofsecrets, BorgHunter, Bork, BostonRed, Brainwad, Brandonrush, BrianHolthouse, Briananthonywaynelee, Brianreed15, Brion VIBBER, Brutaldeluxe, Bryantparnell142, Buaidh, Buddha24, Buddysystem, Bullshark44, Bumm13, Bunterhandy, Burkinaboy, Burn1030, Burroughsks88, Bwintor, C.Fred, CJLL Wright, CMSdline, Cacycle, Calabraxthis, CalicoCatLover, Calmer Waters, Cambrant, Cammc88, CanadianLinuxUser, CanisRufus, Caponer, Capricorn42, Cardsplayer4life, Cburnett, Cchow2, Censusdata, Chairmanriot, ChaosNil, Charlesblackledge, CharlotteWebb, Chicago god, Choster, Chovain, Chowbok, ChrisRuvolo, Chrisdelacruz2000, Chrislk02, Christian List, Chun-hian, Chzz, Circadicrevenge, Circeus, Civil Engineer III, Ckatz, ClamDip, Cmrgcorp, Coffee, Col. Neo, Colonies Chris, Conversion script, Cool Stuff Is Cool, CopperSquare, Corvus cornix, Cottonshirt, Courcelles, Crazy one 55, CrazyC83, Crobertson, Cryptic C62, CutOffTies, Czrisher, D-Day, DARTH SIDIOUS 2, DCEdwards1966, DECOR8Rgirl, DM Andy, DMacks, DRTllbrg, DXRAW, Dabomb87, Dale Arnett, Dalec527, Dan D. Ric, Danger, Daniel, DanielCD, Danny, Danyclement, Dasani, Dausha, Daven200520, David Wahler, DavidSteinle, Dbtfz, Dcornwall, Deathbird909, Decumanus, Deflective, DeltaCreative, Dendodge, DerHexer, Devin Pederson, Diagonalfish, Diannaa, Diocles, Discospinster, DivineIntervention, Djackino, Djharrity, Dlfreem, Dlohcierekim, Docether, Docu, DoubleBlue, Doulos Christos, Download, Dr. Blofeld, Dreamdram, Drsowell, Dspradau, Dufekin, Duja, Dwilso, Dyanega, Dycedarg, Dysepsion, Dysprosia, E23, ESkog, Easter Monkey, Eastlaw, Eco84, Ed g2s, Edderso, Edison, Efmovie, Ekman, El C, ElChupanebre, Elassint, EncycloPetey, Enviroboy, Eoghanacht, Epolk, ErgoSum88, Etisvgxvzxksdjfhj,.gbk;a, Eu.stefan, Euniana, Everyking, Evice, Excirial, Exert, Extreme outdoors, Fakeprof1, Fashionprada333, Fastilysock, Fattyjwoods, Feinoha, Firsfron, Fishing, Flatterworld, Flowanda, Fnlayson, Fogherty V. Tatin, Fomeeko, Footwarrior, Forever Dusk, Foxj, Fred Bradstadt, Frietjes, Fryed-peach, Frymaster, Funandtrvl, Fuzheado, Fæ, GL, Gaius Cornelius, Gcapp1959, GeeZee, Generic Politician, Ghostofarkansas, Giants27, Gihmah, Gil Gamesh, Gilliam, Gimmetrow, Glane23, Glnwalker, Glynner13, Gmatsuda, Goblueaddis, Gobonobo, Gogo Dodo, Golbez, GoldRingChip, Golfcourseairhorn, Good Olfactory, Goodolclint, GorillaWarfare, Got2BRamZ, Gphoto, GraemeL, GraemeMcRae, GregU, Greybear1701, Grim23, Ground Zero, Gruber76, Grunt, GwydionM, Habaneroman, Hadal, Hahagotyou, Hall Monitor, Halmstad, HamburgerRadio, Hammersoft, HangingCurve, Happyfatman021, Harej, Harmil, Harry, Hasek is the best, Hebrew613, HelenSimons, HenkvD, Henrygeorgick, Hereforhomework2, Herrick, HexaChord, HighKing, Highfields, Highway69, Himm, Hmains, HornetMike, Horologium, Howcheng, Hsvbiz, Huntington, I Feel Tired, I only vandalize articles, IAnything21, IEP level 4 Wednesday, II MusLiM HyBRiD II, Ice Cold Beer, Ieplevel4class, InaMaka, Infrogmation, Intro2elephant6, InvertedSaint, InvictaHOG, Iridescent, Isinbill, Ispy1981, J.delanoy, J.kirk, JCarriker, JForget, JFreeman, JNW, JW1805, Jack Cox, Jacob jose, James Kemp, James086, Jason Potter, Java13690, Java7837, Javierito92, JayLilRok06, Jbburt, Jc103089, Jcam, Jcbarr, Jengod, JeremyA, Jesse Viviano, JesseGarrett, JesseW, Jessxenos, JetLover, Jeversol, Jhacob, Jhendin, Jhortman, Jibbajabba, Jim1138, JimIrwin, JimWae, JinJian, Jkatzen, Jmerchant29, JoeSmack, John K, JohnInDC, Jojalozzo, Jojhutton, Jon emmets, Jonnydeluxe, Jorgieporgie, Jose77, Joseph Solis in Australia, Jringer, Jtygs, Juantay, Judeeclare, Juliancolton, Jultemplet, Jumbuck, JustAGal, Juzeris, Jwdoom, KCinDC, KGasso, KI, Kablammo, Kahuroa, Kane5187, Kanonkas, Kazrak, Kazvorpal, Kb3edk, Keith Edkins, Ken Gallager, Ken g6, Keraunoscopia, Kesac, Khoikhoi, Khurg100, King Pickle, King Toadsworth, King of Hearts, Kingpin13, Kintetsubuffalo, Kkarma, Klfinne, Klundarr, Kobalt915m, Kocio, Koyaanis Qatsi, Kozuch, Kralizec!, Kristinpedia, Kthor, Kubigula, Kumioko, Kuru, Kurykh, Kwamikagami, Ky11989, Ky91, L'Aquatique, Lan56, Latash, Lavalette1, Lbbzman, LeaveSleaves, Leuko, Levineps, Lightmouse, LilMane, Lilac Soul, Little Mountain 5, LittleChu542, LonelyPilgrim, Loonymonkey, Looxix, Looziannasuxs, Lord Voldemort, Lottoid, LouI, Ltwin, Luna Santin, Lupo, M C Y 1008, MD87, MJCdetroit, MK8, MONGO, MSJapan, Magioladitis, Magog the Ogre, Manuel Trujillo Berges, MarcoTolo, Marcus9988, Marek69, Martarius, MartinDK, Mastrchf91, Matijap, Matthewrbowker, MattieTK, Mav, McSly, Mcgotime, Mcpusc, Mdd4696, MeStevo, MeltBanana, Meters, Mets501, Michaelthompsonrules, Midnight Green, Mike s, MikeJ9919, MikeLynch, Mikevegas40, Minesweeper, Minguslingus, Minimac, Missar, Misza13, Mjl0509, Moreau36, Mouse is back, Moverton, Mr.grantevans2, Mruffin3, Mrwiki355, Ms.jacob black, Mule Man, Muéro, Mvsmostwnted, Mwanner, Mxn, Myrtlecharlotte, NE2, NEICenergy, NHRHS2010, Nakon, Natalie Erin, NativeForeigner, Natterdawg, NawlinWiki, Nehrams2020, NerdyScienceDude, Neutrality, Neverquick, Niceguyedc, Nick Wilson, Ninjaphobos, Nixeagle, Nkrosse, Nksawyer, Nlu, NotYouHaha, Numbo3, Nurg, Nuujinn, OJH, Oda Mari, Ohnoitsjamie, Ojigiri, Okiefromokla (old), OldRight, Onthegogo, Orphan Wiki, Ortolan88, Otets, Paleok, Pammb, Parkwells, Patrick, Pauly04, Pedro, Perryville12, Pesanserinus, Peter, Peter Karlsen, Peytonio, Pfly, Phale007, Pharaoh of the Wizards, Philip Trueman, Philippe, PhnomPencil, Phoebe, Photolitherland, Phynicen, Pigman, Pigtrino, Pimp998, Pinethicket, Pinikas, Plastikspork, Poeloq, Pokerfacepatrol, Poor Yorick, Popcony, Postdlf, Postpostmod, Prari, Pras, Pras.kota, Premiercolleges, Private Area, Prodego, Pt1234, Pwt898, Qqqqqq, Quadell, Queson, Quidam65, Quintote, Qwyrxian, Qxz, R'n'B, R.T.Gellar, R000t, RBBrittain, RED WHITE AND BLUE11, RVJ, RaCha'ar, Rabbabodrool, Ram-Man, Randomglitter, Raul654, Razorbacks2009, Reach Out to the Truth, RebelAt, Reconfirmer, RedWolf, Reesh, Relevantsus, Rentaferret, Rettetast, Revas, RexNL, Rexmonaco, Rhobite, Rich Farmbrough, Richard David Ramsey, Rick Block, Ricky81682, Rivertorch, Rjensen, Rjwilmsi, Rlhowk, Rlove, Rmcclen, Rmcdou1, Rob Hooft, Robbie Mac, Robdurbar, Robertablake, Robertb-dc, Robyonone, Rocastelo, Rokfaith, Romanm, Romeisburning, Root Beers, Roozbeh, Rorschach, Rpyle731, Rrburke, Rreagan007, Ruby Rose, Ruhrfisch, Rushadthomas, Sallyorbecky, SaltyBoatr, Sam hogg, Sango123, Saopaulo1, Sardanaphalus, SatyrTN, Savh, Scanlan, SchfiftyThree, SchirmerPower, Schmiteye, Schnubble, Schzmo, Sciurinæ, Scott5114, ScottMainwaring, ScottSteiner, Scoutcalvert, Seanmagill, Seaphoto, Secret (renamed), Sedna10387, Sellyme, Senarclens, Seveikath, Sevela.p, Sewhitlock, Sfan00 IMG, Sfmontyo, Sgr2000, ShelfSkewed, Shereth, SilkTork, Silverxxx, SinjinRJ, Skillz187, Skinsmoke, Skizzik, Sligocki, Slthree, Smallbones, Smalljim, Smallman12q, Snideology, Snowdog, Softballchick23, Solipsist, Some jerk on the Internet, SoulMeetsBody, SpaceFlight89, Specs112, Speedoflight, Spiderman4876924578, Spqr369, Springflight, Spydrlink, St3class, StaticGull, Staxringold, Stern, Stevewonder2, StoptheDatabaseState, Sugarrose, SupaStarGirl, Supercoop, Superm401, SusieMcgangbeng, Svetovid, Svgalbertian, SwampStomp, Swampfox117, Sweetmoose6, Swerdnaneb, Swollib, TUF-KAT, Tarkan1st, Tarquin, Tassedethe, Tater9466, Tbhotch, Techbonefrombama, Tedickey, Template namespace initialisation script, Tesscass, TexasAndroid, ThaddeusB, Thatguyflint, The Earwig, The Epopt, The J-Flow, The Obento Musubi, The Thing That Should Not Be, The Transhumanist, The Universe Is Cool, The chavi, The stuart, TheOnewithblueeyes, TheRingess, TheTito, Thedude2000, Thejamie024, Thermal0xidizer, Thesouthernhistorian45, Tholly, Thrane, Thunderstruck401, Tide rolls, Tidying Up, Timneu22, Tobby72, Tom93845790231, Tommy2010, TommyBoy, Topbanana, Tpbradbury, Tracer9999, Transity, Trebor, Trekphiler, Treyt021, Trianglesquaredot, Tw33tee, TwinCityIL, Twistingthenightaway, Ucabear1982, Ucabears2013, Ucucha, Ufwuct, Ukexpat, Ulmanor, Upwardbound2010, Urhixidur, Utcursch, Ute in DC, Vegaswikian, Velvet elvis81, Velvetron, Verloren, Versus22, VeryVerily, Vina, VolatileChemical, Vox Rationis, Vpuliva, Vsmith, W 246g7ya8j, WPANI, Wajay 47, Waltpohl, Wangi, Wapcaplet, Wars, WarthogDemon, Wayne Slam, Weatherman78, Wesley M. Curtus, West.andrew.g, Whereizben, WhisperToMe, White Shadows, Who then was a gentleman?, Why Not A Duck, WiccaIrish, WikHead, Wiki alf, WikiPuppies, Wikieditor06, Wikipelli, Wikisidd, Will Beback, WillC, Willking1979, Wilybadger, Wittylama, Wknight94, Wmahan, Wmmccall210@hotmail.com, Woohookitty, Wrtg1320tth925, Wtmitchell, Wwb, Wæng, XJamRastafire, Xezbeth, Xsixwings, Xyzzyva, Y, Yamamoto Ichiro, Yath, Ybbor, Yekrats,

Image Sources, Licenses and Contributors

File:Arkansas 204.svg *Source*: http://en.wikipedia.org/w/index.php?title=File:Arkansas_204.svg *License*: unknown *Contributors*: Common Good, Ltljltlj

File:US 71B.svg *Source*: http://en.wikipedia.org/w/index.php?title=File:US_71B.svg *License*: unknown *Contributors*: User:Fredddie

File:Map of Arkansas highlighting Benton County.svg *Source*: http://en.wikipedia.org/w/index.php?title=File:Map_of_Arkansas_highlighting_Benton_County.svg *License*: unknown *Contributors*: User:Dbenbenn

File:Map of USA AR.svg *Source*: http://en.wikipedia.org/w/index.php?title=File:Map_of_USA_AR.svg *License*: unknown *Contributors*: User:Huebi

Image:I-540 (AR).svg *Source*: http://en.wikipedia.org/w/index.php?title=File:I-540_(AR).svg *License*: unknown *Contributors*: Ltljltlj, Xnatedawgx

Image:US 62 (1961).svg *Source*: http://en.wikipedia.org/w/index.php?title=File:US_62_(1961).svg *License*: unknown *Contributors*: User:Fredddie, User:SPUI, User:Scott5114

Image:US 71 (1961).svg *Source*: http://en.wikipedia.org/w/index.php?title=File:US_71_(1961).svg *License*: unknown *Contributors*: User:Fredddie, User:SPUI, User:Scott5114

Image:US 412.svg *Source*: http://en.wikipedia.org/w/index.php?title=File:US_412.svg *License*: unknown *Contributors*: User:Vishwin60

Image:Arkansas 12.svg *Source*: http://en.wikipedia.org/w/index.php?title=File:Arkansas_12.svg *License*: unknown *Contributors*: Ltljltlj, Sevela.p

Image:Arkansas 16.svg *Source*: http://en.wikipedia.org/w/index.php?title=File:Arkansas_16.svg *License*: unknown *Contributors*: Ltljltlj, Sevela.p

Image:Arkansas 43.svg *Source*: http://en.wikipedia.org/w/index.php?title=File:Arkansas_43.svg *License*: unknown *Contributors*: Ltljltlj, Sevela.p

Image:Arkansas 59.svg *Source*: http://en.wikipedia.org/w/index.php?title=File:Arkansas_59.svg *License*: unknown *Contributors*: Ltljltlj, Sevela.p

Image:Arkansas 72.svg *Source*: http://en.wikipedia.org/w/index.php?title=File:Arkansas_72.svg *License*: unknown *Contributors*: Ltljltlj, Sevela.p

Image:Arkansas 94.svg *Source*: http://en.wikipedia.org/w/index.php?title=File:Arkansas_94.svg *License*: unknown *Contributors*: Ltljltlj, Sevela.p

Image:Arkansas 102.svg *Source*: http://en.wikipedia.org/w/index.php?title=File:Arkansas_102.svg *License*: unknown *Contributors*: Ltljltlj, Sevela.p

Image:USA Benton County, Arkansas age pyramid.svg *Source*: http://en.wikipedia.org/w/index.php?title=File:USA_Benton_County,_Arkansas_age_pyramid.svg *License*: unknown *Contributors*: user:WarX

File:Benton County Arkansas 2010 Township Map small.jpg *Source*: http://en.wikipedia.org/w/index.php?title=File:Benton_County_Arkansas_2010_Township_Map_small.jpg *License*: unknown *Contributors*: Yx7791

File:Benton_County_Arkansas_Incorporated_and_Unincorporated_areas_Siloam_Springs_Highlighted.svg *Source*: http://en.wikipedia.org/w/index.php?title=File:Benton_County_Arkansas_Incorporated_and_Unincorporated_areas_Siloam_Springs_Highlighted.svg *License*: unknown *Contributors*: Arkyan

Image:Siloam Springs Arkansas Fountains.JPG *Source*: http://en.wikipedia.org/w/index.php?title=File:Siloam_Springs_Arkansas_Fountains.JPG *License*: unknown *Contributors*: User:Tim Morgan

Image:XNA_logo.png *Source*: http://en.wikipedia.org/w/index.php?title=File:XNA_logo.png *License*: unknown *Contributors*: Zyxw

file:USA Arkansas location map.svg *Source*: http://en.wikipedia.org/w/index.php?title=File:USA_Arkansas_location_map.svg *License*: unknown *Contributors*: User:Alexrk2

File:Airplane_silhouette.svg *Source*: http://en.wikipedia.org/w/index.php?title=File:Airplane_silhouette.svg *License*: unknown *Contributors*: User:McSush

Image:KXNA Route Map.png *Source*: http://en.wikipedia.org/w/index.php?title=File:KXNA_Route_Map.png *License*: unknown *Contributors*: D.c.camero

File:US 71.svg *Source*: http://en.wikipedia.org/w/index.php?title=File:US_71.svg *License*: unknown *Contributors*: Bidgee, SPUI, Xnatedawgx, 2 anonymous edits

File:By-pass plate.svg *Source*: http://en.wikipedia.org/w/index.php?title=File:By-pass_plate.svg *License*: unknown *Contributors*: Homefryes, Ltljltlj, RTCNCA, SPUI

File:Business plate.svg *Source*: http://en.wikipedia.org/w/index.php?title=File:Business_plate.svg *License*: unknown *Contributors*: Homefryes, Ltljltlj, RTCNCA, SPUI

File:Southern terminus of US 71B, Fayetteville, AR.jpg *Source*: http://en.wikipedia.org/w/index.php?title=File:Southern_terminus_of_US_71B,_Fayetteville,_AR.jpg *License*: unknown *Contributors*: User:Brandonrush

File:School Ave intersecting MLK Blvd, Fayetteville, AR.jpg *Source*: http://en.wikipedia.org/w/index.php?title=File:School_Ave_intersecting_MLK_Blvd,_Fayetteville,_AR.jpg *License*: unknown *Contributors*: User:Brandonrush

File:US Route 71B intersects US 412, Springdale, AR.jpg *Source*: http://en.wikipedia.org/w/index.php?title=File:US_Route_71B_intersects_US_412,_Springdale,_AR.jpg *License*: unknown *Contributors*: User:Brandonrush

File:US 71 (1961).svg *Source*: http://en.wikipedia.org/w/index.php?title=File:US_71_(1961).svg *License*: unknown *Contributors*: User:Fredddie, User:SPUI, User:Scott5114

File:Arkansas 16.svg *Source*: http://en.wikipedia.org/w/index.php?title=File:Arkansas_16.svg *License*: unknown *Contributors*: Ltljltlj, Sevela.p

File:Arkansas 180.svg *Source*: http://en.wikipedia.org/w/index.php?title=File:Arkansas_180.svg *License*: unknown *Contributors*: Ltljltlj, Sevela.p

File:Arkansas 45.svg *Source*: http://en.wikipedia.org/w/index.php?title=File:Arkansas_45.svg *License*: unknown *Contributors*: Ltljltlj, Luigi Chiesa

File:I-540 (AR) Metric.svg *Source*: http://en.wikipedia.org/w/index.php?title=File:I-540_(AR)_Metric.svg *License*: unknown *Contributors*: Ltljltlj

File:US 62 (1961).svg *Source*: http://en.wikipedia.org/w/index.php?title=File:US_62_(1961).svg *License*: unknown *Contributors*: User:Fredddie, User:SPUI, User:Scott5114

File:US 412 (AR).svg *Source*: http://en.wikipedia.org/w/index.php?title=File:US_412_(AR).svg *License*: unknown *Contributors*: User:Fredddie

File:Arkansas 264.svg *Source*: http://en.wikipedia.org/w/index.php?title=File:Arkansas_264.svg *License*: unknown *Contributors*: Common Good, Ltljltlj

File:Arkansas 94.svg *Source*: http://en.wikipedia.org/w/index.php?title=File:Arkansas_94.svg *License*: unknown *Contributors*: Ltljltlj, Sevela.p

File:Arkansas 12.svg *Source*: http://en.wikipedia.org/w/index.php?title=File:Arkansas_12.svg *License*: unknown *Contributors*: Ltljltlj, Sevela.p

File:Arkansas 112.svg *Source*: http://en.wikipedia.org/w/index.php?title=File:Arkansas_112.svg *License*: unknown *Contributors*: Ltljltlj, Sevela.p

File:Arkansas 102.svg *Source*: http://en.wikipedia.org/w/index.php?title=File:Arkansas_102.svg *License*: unknown *Contributors*: Ltljltlj, Sevela.p

File:Arkansas 72.svg *Source*: http://en.wikipedia.org/w/index.php?title=File:Arkansas_72.svg *License*: unknown *Contributors*: Ltljltlj, Sevela.p

File:US 71 (IA).svg *Source*: http://en.wikipedia.org/w/index.php?title=File:US_71_(IA).svg *License*: unknown *Contributors*: User:Fredddie

File:Iowa 2.svg *Source*: http://en.wikipedia.org/w/index.php?title=File:Iowa_2.svg *License*: unknown *Contributors*: Scott Onson

File:No image wide.svg *Source*: http://en.wikipedia.org/w/index.php?title=File:No_image_wide.svg *License*: unknown *Contributors*: SPUI, TwinsMetsFan

File:Iowa 7.svg *Source*: http://en.wikipedia.org/w/index.php?title=File:Iowa_7.svg *License*: unknown *Contributors*: Scott Onson

File:Alternate plate.svg *Source*: http://en.wikipedia.org/w/index.php?title=File:Alternate_plate.svg *License*: unknown *Contributors*: Homefryes, Ltljltlj, RTCNCA, SPUI

Image:ALT US 71.JPG *Source*: http://en.wikipedia.org/w/index.php?title=File:ALT_US_71.JPG *License*: unknown *Contributors*: Original uploader was Rt66lt at en.wikipedia

file:Seal of the Arkansas State Highway and Transportation Department.svg *Source*: http://en.wikipedia.org/w/index.php?title=File:Seal_of_the_Arkansas_State_Highway_and_Transportation_Department.svg *License*: unknown *Contributors*: Government of Arkansas

File:Decrease2.svg *Source*: http://en.wikipedia.org/w/index.php?title=File:Decrease2.svg *License*: unknown *Contributors*: User:Sarang

File:Flag of Arkansas.svg *Source*: http://en.wikipedia.org/w/index.php?title=File:Flag_of_Arkansas.svg *License*: unknown *Contributors*: Awg1010, Dbenbenn, Dzordzm, Fry1989, Himasaram, Homo lupus, Nightstallion, Serinde, Smooth O, Vantey, Zscout370, 1 anonymous edits

File:Seal of Arkansas.svg *Source*: http://en.wikipedia.org/w/index.php?title=File:Seal_of_Arkansas.svg *License*: unknown *Contributors*: State of Arkansas

File:Arkansas in United States.svg *Source*: http://en.wikipedia.org/w/index.php?title=File:Arkansas_in_United_States.svg *License*: unknown *Contributors*: TUBS

Image:Speakerlink.svg *Source*: http://en.wikipedia.org/w/index.php?title=File:Speakerlink.svg *License*: unknown *Contributors*: Woodstone. Original uploader was Woodstone at en.wikipedia

File:Petit Jean State Park view.jpg *Source*: http://en.wikipedia.org/w/index.php?title=File:Petit_Jean_State_Park_view.jpg *License*: unknown *Contributors*: Brandonrush (talk). Original uploader was Brandonrush at en.wikipedia

File:BuffaloRiver.jpg *Source*: http://en.wikipedia.org/w/index.php?title=File:BuffaloRiver.jpg *License*: unknown *Contributors*: Original uploader was Vsmith at en.wikipedia

File:Flatsidewildernesssunset1a.jpg *Source*: http://en.wikipedia.org/w/index.php?title=File:Flatsidewildernesssunset1a.jpg *License*: unknown *Contributors*: User:Dismalhiker

File:Wife and children of a sharecropper in Washington County, Arkansas - NARA - 195845.tif *Source*: http://en.wikipedia.org/w/index.php?title=File:Wife_and_children_of_a_sharecropper_in_Washington_County,_Arkansas_-_NARA_-_195845.tif *License*: unknown *Contributors*: Djembayz, Origamiemensch

Printed by Books on Demand GmbH, Norderstedt / Germany

dem immer autoritärer auftretenden Erdogan. Als Antwort öffnete Erdogan die Türkei zum Durch- und Weiterzug für die Flüchtlinge und zwang damit die Europäer zum Einlenken.

Als der Flüchtlingsansturm auf Europa traf, war Europa unvorbereitet. Die Politik floh in allgemeine Erklärungsmuster, um ihr bisheriges Versagen zu überdecken. Eine Flüchtlingspolitik, die ihren Namen verdient, besaß Europa nicht und hat sie bis heute nicht entwickelt. Angela Merkel antwortete mit der *Willkommenskultur* und spielte damit Erdogan noch in die Hände.

Mit dem Vertrag vom 18. März 2016 zwischen der EU und der Türkei ebbt der Flüchtlingsstrom aus der Türkei ab. Die Türkei hat den freien Strom der Flüchtlinge durch ihr Land beendet. Erdogan hatte seine politischen Ziele erreicht. Er hatte sich aus der politischen Isolierung befreit.

Insbesondere in Deutschland hat die Flüchtlingskrise schwerwiegende Folgen hinterlassen. Von den Flüchtlingen, die 2015 in der EU Schutz gesucht haben, hat Deutschland mit einer Million Flüchtlingen den Großteil aufgenommen. Angela Merkel hatte von den EU Partnern verlangt, dass sie einen Teil der Flüchtlingslasten übernähmen. Sie konnte sich mit ihrer Forderung nicht durchsetzen. Ein Teil der Mitgliedsländer wie die osteuropäischen Staaten lehnt jede Beteiligung ab, der größere Teil wie z.B. Frankreich war nur zur Übernahme kleinerer Kontingente bereit. Der Versuch, eine Aufteilung zwangsweise durchzusetzen, versandete. Er hinterließ erhebliche Spannungen innerhalb der EU.

Angela Merkel hatte ihre *Willkommenskultur* als politisch-moralisches Axiom vertreten, dem sich nicht nur Deutschland, sondern auch die übrigen Europäer unterzuordnen hätten. Diese Staaten verweigerten sich. Sie wollen in ihrer Souveränität selbst entscheiden, wen und wie viele Menschen sie zu sich lassen. Auch in Deutschland entwickelte sich Widerstand. Er fand neben der bayerischen CSU sein Sprachrohr in einer erstarkenden AfD.